Free Will

Free Will

*Philosophers and Neuroscientists
in Conversation*

Edited by

URI MAOZ
AND
WALTER SINNOTT-ARMSTRONG

OXFORD
UNIVERSITY PRESS

OXFORD
UNIVERSITY PRESS

Oxford University Press is a department of the University of Oxford. It furthers
the University's objective of excellence in research, scholarship, and education
by publishing worldwide. Oxford is a registered trade mark of Oxford University
Press in the UK and certain other countries.

Published in the United States of America by Oxford University Press
198 Madison Avenue, New York, NY 10016, United States of America.

Library of Congress Cataloging-in-Publication Data
Names: Maoz, Uri, editor. | Sinnott-Armstrong, Walter, 1955– editor.
Title: Free will : philosophers and neuroscientists in conversation /
edited by Uri Maoz and Walter Sinnott-Armstrong, Duke University.
Description: New York, NY, United States of America :
Oxford University Press, [2022] |
Includes bibliographical references and index.
Identifiers: LCCN 2021023162 | ISBN 9780197572160 (paperback) |
ISBN 9780197572153 (hardback) | ISBN 9780197572184 (epub)
Subjects: LCSH: Free will and determinism—Miscellanea. | Philosophy and
science—Miscellanea. | Medicine—Philosophy—Miscellanea. |
Neurosciences—Miscellanea.
Classification: LCC BJ1461 .F7545 2021 | DDC 123/.5—dc23
LC record available at https://lccn.loc.gov/2021023162

DOI: 10.1093/oso/9780197572153.001.0001

From Uri:
To Tatiana and Michael for their support and encouragement;
to Beata, Tommy, and Danny for their enthusiasm and patience;
to Paul Gluck for his infectious love of science;
as if they had a choice . . .

From Walter:
To David, Don, Ed, Eric, Orin, Patrice, Randy,
and all of my other golfing buddies at Duke,
for freely choosing to play with me.

Contents

SECTION III: QUESTIONS ABOUT SCIENTIFIC EVIDENCE

SECTION IV: QUESTIONS ABOUT CONSCIOUSNESS

SECTION V: QUESTIONS ABOUT RESPONSIBILITY AND REASONS-RESPONSIVENESS

PART II: QUESTIONS FROM PHILOSOPHERS FOR NEUROSCIENTISTS

SECTION I: QUESTIONS ABOUT WILL

SECTION II: QUESTIONS ABOUT INTENTION

SECTION III: QUESTIONS ABOUT CONSCIOUSNESS

SECTION IV: QUESTIONS ABOUT NEUROSCIENCE METHODS

Preface

Walter Sinnott-Armstrong and Uri Maoz

Philosophers have debated the extent of human free will for millennia. Over recent decades, neuroscientists have increasingly studied and elucidated human actions and decisions. Despite some initial reluctance, both fields have begun to see value in the contribution of the other. Many neuroscientists realized the importance of greater conceptual clarity and precision for high-level concepts like volition and consciousness in defining research questions and interpreting empirical results. And many philosophers now view several important questions related to volition and control as empirically tractable.

Collaboration across these disciplinary boundaries is anything but easy because neuroscientists and philosophers have different scholarly traditions, methods of investigation and discourse, and even career incentives. One problem is that professional philosophers too often stick with questions that only academic philosophers care about, and neuroscientists too often focus on technical details and lose the forest for the trees. Another stubborn barrier is language, since philosophers and neuroscientists often refer to somewhat different concepts of will, decision, intention, action, freedom, and so on. Yet another barrier is time, for busy philosophers and neuroscientists find it difficult to keep up with the rapid, relevant developments in another large field.

If they are to overcome these obstacles, philosophers and neuroscientists need to engage in detailed conversations and thereby learn to talk to—rather than past—each other. A good way to begin is to identify the most important questions to answer. This book began with a suggestion by one of the editors (Sinnott-Armstrong) that nine philosophers and eight neuroscientists, who come from seven countries on four continents, should ask each other the questions they would most like the other discipline to answer. We gathered over 30 questions from each discipline. After some editing, the philosophers voted for their top choices, and so did the neuroscientists. We ended up with 15 questions from each group for the other group to answer.

The editors then assigned the questions to the most suitable experts within this group of 17 scholars. The answers to those 30 questions were edited and revised, sometimes repeatedly, in light of comments by the editors, their research assistants, and authors of other chapters. Then both philosophers and neuroscientists were given the opportunity to ask follow-up questions about each of the 30 answers. Finally, with guidance from the editors, the authors chose follow-up questions and responded to them, leading to another round of editing and revisions.

This book therefore contains 30 direct, bidirectional exchanges between neuroscientists and philosophers that focus on the most critical questions in the neurophilosophy of free will. It thus mimics a lively, interdisciplinary conference, where experts answer questions and follow-up questions from the other field, helping each discipline to understand how the other thinks and works. Each chapter is concise and accessible to non-experts—free from disciplinary jargon and highly technical details—but also accurate enough to satisfy experts. To further assist non-experts, we included a glossary, an annotated bibliography, and some relevant brain maps.

The resulting collection should be useful to anyone who wants to get up to speed on the most fundamental issues in the rising field of the neurophilosophy of free will. It will interest experts in each field who want to learn about the other discipline, students in courses on a host of related topics, and lay readers who are fascinated by these profound issues. It should also stimulate and provide a model for future cross-disciplinary exchanges, because other fields can also benefit from similar conversations.

The questions that philosophers answered were divided into five topics: will, freedom, scientific evidence, consciousness, and responsibility. Those answered by neuroscientists were divided into four topics: will, intention, consciousness, and neuroscience methods. Readers, thus, have a choice. They may read the chapters in the order presented here—first the chapters by philosophers and then the chapters by neuroscientists—or they may read the answers by neuroscientists before the answers by philosophers.

It is equally feasible to read the sets of chapters topic by topic, such as by reading the chapters by philosophers on will followed by the chapters by neuroscientists on will and intention, then the chapters by philosophers and neuroscientists on consciousness, and then the chapters by philosophers on scientific evidence, followed by the chapters by neuroscientists on neuroscience methods, and ending with the chapters by philosophers on freedom and responsibility. Of course, there are many other ways to use this material, depending on one's background and interests.

These concise chapters are not intended to provide the final word on any of these complex questions. For those who want to explore the topics more deeply, the book also contains an annotated bibliography with suggestions for further reading as well as a platform to help readers locate the cited areas of the brain that are important to the field.

This book is a product of a large, interdisciplinary project on the neurophilosophy of free will, which is sponsored by the John Templeton Foundation and the Fetzer Institute under the leadership of one of the editors (Maoz). We would like to thank these sponsors for their vision and trust, without which this collaboration would not have been possible. We note that the opinions expressed in this publication are those of the authors and do not necessarily reflect the views of the John Templeton Foundation or the Fetzer Institute. The book was also supported by the Fetzer Franklin Fund of the John E. Fetzer Memorial Trust, which allowed us to hire research assistants—Deniz Arıtürk, Amber Hopkins, and Claire Simmons—who were invaluable throughout this process. It also enabled us to hire graphics artist Natalie Nichols, who diligently and patiently created and modified the brain maps. Furthermore, we wish to thank the Federico and Elvia Faggin Foundation for their support. We also thank Marta Napiorkowska for comments on an earlier version of this manuscript. Last, but certainly not least, we thank the entire group of 17 philosophers and neuroscientists, along with their lab members, who put this book together. More than an edited volume, this is a book with many authors.

List of Contributors

Amir, Yoni (Tel Aviv University)

Arıtürk, Deniz (Duke University)

Bayne, Tim (Monash University)

Block, Ned (New York University)

Bold, Jye lyn (Chapman University)

Dominik, Tomáš (Chapman University)

Gavenas, Jake (Chapman University)

Haggard, Patrick (University College London)

Hall, Jonathan (University of Edinburgh)

Hallett, Mark (National Institute of Neurological Disorders and Stroke, National Institutes of Health)

Haynes, John-Dylan (Charité—Universitätsmedizin Berlin)

Hieronymi, Pamela (UCLA)

Hopkins, Amber (Chapman University)

Jeay-Bizot, Lucas (Chapman University)

Kreiman, Gabriel (Harvard Medical School)

Lee, Sae Jin (National Institute of Neurological Disorders and Stroke, National Institutes of Health)

Liang, Dehua (Andy) (Chapman University)

Liljenström, Hans (Agora for Biosystems, Swedish University of Agricultural Sciences)

Lynch, Tierra (Dartmouth College)

Maoz, Uri (Chapman University, UCLA, Caltech)

Mele, Alfred R. (Florida State University)

Mudrik, Liad (Tel Aviv University)

Nahmias, Eddy (Georgia State University)

O'Connor, Timothy (Indiana University, Bloomington)

Parés-Pujolràs, Elisabeth (University College London)

Roskies, Adina L. (Dartmouth College)

Rotbain, Tzachi (Tel Aviv University)

Schurger, Aaron (Chapman University and INSERM)

Seghezzi, Silvia (University of Milano-Bicocca)

Silverstein, David N. (Agora for Biosystems)

Simmons, Claire (Duke University)

Sinnott-Armstrong, Walter (Duke University)

Triggiani, Antonio Ivano (National Institute of Neurological Disorders and Stroke, National Institutes of Health)

Vierkant, Tillmann (University of Edinburgh)

Wong, Sook Mun (Alice) (Chapman University)

Yaffe, Gideon (Yale University)

Zhang, Mengmi (Harvard Medical School)

PART I

QUESTIONS FROM NEUROSCIENTISTS FOR PHILOSOPHERS

PART I, SECTION I
QUESTIONS ABOUT WILL

1

What is an intention?

Gideon Yaffe

The dominant position among philosophers is that intentions are mental representations of future (or perhaps present) states which play distinctive causal roles in human psychology and which are governed by distinctive rational requirements. To have an intention is to have a thought about a future action, where the thought has certain distinctive effects on behavior and reasoning and is subject to distinctive rational requirements.[1]

Not all mental representations of the future are intentions, for not all representations of the future either play the same causal roles as intentions or are subject to the same rational requirements. Intentions differ, for instance, from beliefs, or predictions about the future, in what philosophers sometimes call "direction of fit." The important philosopher Elizabeth Anscombe (2000) illustrated this idea with an example: A goes to the grocery store with a list and fills his cart to match the list. B spies on A, writing down each thing that A puts in his cart. Now imagine that there is an orange in the cart, but the word "orange" appears on neither A's nor B's lists. A should fix the problem by removing the orange from his cart. B should fix the problem, by

---

[1] According to a prominent minority view, represented in the work of Michael Thompson (2012), whether it is true that a person has an intention, for instance, to travel to New York tomorrow is a matter of whether certain sorts of questions are appropriately asked of her. Is it appropriate, for instance, to expect her to identify something desirable about being in New York tomorrow in answer to the question "Why are you going to New York tomorrow?" Since what questions it is appropriate to ask, and what answers it is appropriate to expect, are social matters and a function of much more than is in the head of any one person, on this view, intentions are not, as on the dominant view, individual thoughts about future actions. However, under this minority view, intentions are also not the kinds of thing that one could directly investigate neuroscientifically. To do so would be like using neuroscience to study government; it's not the kind of thing that can be investigated directly by measuring the brain.

Gideon Yaffe, *What is an intention?* In: *Free Will.* Edited by: Uri Maoz and Walter Sinnott-Armstrong, Oxford University Press. © Oxford University Press 2022. DOI: 10.1093/oso/9780197572153.003.0001

contrast, by adding the word "orange" to his list. If A's list functions as it ought to, A will change the world to match it. By contrast, if B's list functions well, B will change the list to match the world. The two lists have different "directions of fit." A's list has what you might call "world-to-list" direction of fit; its job is to make the world match the list. B's list has "list-to-world" direction of fit; its job is to match the world as it is.

Our minds function well when our beliefs change to match the world. You believe you will be in New York tomorrow, but learning that your flight is canceled you change your belief, concluding that you will be in London instead. Beliefs have "mind-to-world" direction of fit; they are like the spy's list. But intentions work the opposite way. Thanks to your intention to be in New York tomorrow, you take steps to make it so, by buying plane tickets, for instance. Our intentions function as they ought, that is, not when they change to match the world but when the world changes to match them. They have "world-to-mind" direction of fit; they are like the shopper's list.

However, intentions differ even from other representations of the future that share their direction of fit.[2] Intentions, for instance, are different from desires, which also have world-to-mind direction of fit. If you want to have healthy teeth but do not want to go to the dentist, you are fully rational; healthy teeth are important, and trips to the dentist are no fun. But if you intend to have healthy teeth, and you believe that going to the dentist is the only way to have healthy teeth, but you intend also not to go to the dentist, then you are irrational. Intentions are subject to demands of rational consistency that do not govern desires. This is one of the ways in which intentions are subject to distinctive requirements of rationality: rationality requires that they be consistent with one another, and with one's beliefs, in ways that desires need not be.[3]

---

[2]   In an important book, the philosopher Michael Bratman (1987) pioneered the effort to map the many ways in which intentions differ from other mental states that share their direction of fit. While disagreements abound about exactly what effects on behavior and reasoning intentions have, and what norms of rationality govern them, Bratman's basic idea that intentions are distinctive in light of such effects and norms is shared by most philosophers.

[3]   See Chapter 23 by Hopkins & Maoz on the neural correlates of beliefs and desires.

Further, intentions have effects on behavior and on reasoning that are importantly different from the effects of desires and other states with world-to-mind direction of fit. Intentions resist revision and reconsideration after they are formed, and this resistance helps us to perform complex tasks, for instance. Say you spend several minutes thinking through all the options and form the intention to have spaghetti for dinner. Hours later, you find yourself at the store. It is because you intend to have spaghetti that you do not waste time now, at the grocery store, rethinking the question of what to have for dinner. Intentions settle matters for practical purposes in ways that other mental states, such as mere wishes, do not, and so they save us valuable time that would otherwise need to be spent deliberating anew. This is an example of a distinctive effect of intention on reasoning: they cut short, or prevent, repeated reasoning of the sort that led to their formation. They also prompt forms of reasoning that other mental states do not. Thanks to the intention to have spaghetti for dinner, you think through the question of whether or not to go to the store, a question that you would not have bothered thinking about had you not formed the intention.

In addition to their effects on reasoning, intentions have distinctive effects on behavior that provide for useful forms of coordination both with one's present, past, and future selves and with other people. For instance, if you form the intention to go to New York tomorrow, then that intention might cause you to make a fancy dinner reservation for tomorrow night in New York, putting down a deposit. The money is worth paying because you expect to be there, and you expect to be there because you intend to be there. Further, the chef expects you to be there and so takes steps to bake you a celebratory cake. It is not just you who relies on your future self to behave in a certain way, given that your present self has an intention, but also other people, if they know your intention. Notice that for all this you still might desire to eat tomorrow night not in New York but at a restaurant in Paris. Still, it would be a mistake to put down a deposit there, and it would be a mistake for the Paris chef, even if he knows of your desire, to bake you a cake, since the fact that you have this desire gives no reason for you or anyone else to expect you to be in Paris tomorrow night, especially when you intend to be in New York. Intentions tend to cause the future actions they represent, whereas mere desires without intentions often leave us hoping but not acting.

Intentions, in short, are adaptive. They help us to achieve complex goals by helping us to use our scarce reasoning resources efficiently; to coordinate our behavior with our past and future selves, as well as with other people; and to exploit our rational tendencies to revise them in ways that will help us to further our goals. Mental states that are not adaptive in these distinct ways—although perhaps adaptive in others— are not the appropriate object of inquiry for those who study intention.

Follow-Up Questions

Walter Sinnott-Armstrong

You wrote, "To have an intention is to have a thought about a future action, where the thought has certain distinctive effects on behavior and reasoning and is subject to distinctive rational requirements." In contrast, Parés-Pujolràs and Haggard conclude, "[I]ntentional actions can be defined as actions that follow from an agent's goals and are identified as such by the agent herself."[4] Their definition does not mention any "distinctive rational requirements." Does that mean that they as neuroscientists are talking about different things than philosophers when they both discuss intentions? In particular, neuroscientists often distinguish distal, proximate, and motor intentions.[5] Do these different kinds of intentions all fit into your definition? How can motor intentions be subject to "distinctive rational requirements"?

Silvia Seghezzi and Patrick Haggard

In this answer, you argue that intentions differ from other representations of the future—like beliefs—in the "direction of

---

[4] See Chapter 21 by Parés-Pujolràs & Haggard on intentional actions.
[5] See Chapter 21 by Parés-Pujolràs & Haggard on intentional actions and Chapter 22 by Haynes on evidence that intentions are represented in the brain.

fit": intentions have "world-to-mind" direction of fit, in the sense that our intentions function as they ought when the world changes to match them and not vice versa. But how do intentions differ from other "world-to-mind" states? Could the key difference between intentions and other representations of the future lie in the fact that intentions directly imply action, while desires do not? Can a desire become an intention if we add to a desire an "element of activeness"?

Amber Hopkins

Parés-Pujolràs and Haggard refer to a main "aspect of intentional action: the fact that people (except for some interesting pathological cases) typically *know* whether their actions are intentional or not, and in what sense they are intentional."[6] But you do not mention the subjective feeling in your answer. How important do you think consciousness is for intentions?

Reply to Follow-Up Questions

Gideon Yaffe

Terms like "proximal," "distal," and "motor intention" refer to classes of intentions distinguished by their content—by, that is, what they are intentions to do. If you intend to do something now or in the immediate future, your intention is "proximal"; if in the more distant future, then it is "distal"; if it is an intention to move your body, then it is a "motor intention." While it is easiest to identify the requirements of rationality that apply to distal intentions, those requirements apply also to both proximal and motor intentions. Someone, for instance, who intends to hold his hands still and also intends to clap his hands right now is guilty of a form of irrationality, assuming that he believes that

---

[6] See Chapter 21 by Parés-Pujolràs & Haggard on intentional actions.

he cannot do both. He would not be similarly irrational were he merely to desire to do both things.

There are, however, two not uncommon usages of the term "motor intention" under which it does not refer to an intention at all. Under the first, the term refers to neural activity, such as neural activity in the motor cortex, that causes bodily motion. Since the neural activity in question is not claimed to be a representation of an action, such as a bodily motion, nor is it understood to be governed by any norms of rationality of the sort that govern intention, nor is it claimed even to be a thought when the term is used this way, the term applies to something even if that to which it refers is not an intention at all. When neuroscientists use the term in this way, they speak past philosophers. In fact, they do so even if they use the term in a second common way to refer to *the thought* that is underlain by the neural activity in question without requiring that the thought must meet the distinctive criteria for intention in order for the term to apply.

It is primarily the norms of rationality that apply which distinguish intentions from other mental states with world-to-mind direction of fit. Intentions do have distinctive behavioral effects—they often cause deliberation about how to satisfy them, and they often cause action that matches their content, for instance; they also often *prevent* deliberation about courses of conduct inconsistent with them. But it is because it is possible to have an intention that one never acts on in any direct or even indirect way that we know that behavioral effects are not what distinguish intentions from other world-to-mind mental states like desires. As with desires, it is possible to intend to build a birdhouse but never deliberate about it and never take any steps toward doing it; still, even then, you do have the intention.

That effects of intention are not criterial for their possession is easily overlooked in part because the most plausible reason for thinking that intentions are governed by norms of rationality is that they ordinarily have distinctive behavioral effects that help us to achieve our goals, provided that they conform to certain rules. For instance, intentions are useful to us in part because they tend to cause the acts they represent. When you form an intention to X, you increase the probability that you will X; when X is something that will be helpful to you, then it is useful to form an intention to X. But intentions are not useful in this way when they conflict with one another, relative to our beliefs. If you form an intention to X and also form an intention to not-X, then it is no more likely that you will X than if you never formed any intentions. And this is, plausibly, why it is irrational to have conflicting intentions; they stop being useful when they conflict and they are useful because of the kinds of effects they ordinarily have. Still, what makes an intention an intention are the norms of rationality that apply to it even though it is also true that those norms apply thanks to the useful effects that intentions ordinarily, but not invariably, have.

Notice that under plausible accounts of desires' functionality—accounts of what purpose desires serve—they need not be consistent with our beliefs in order to be useful to us. For instance, desires draw our attention to aspects of objects and actions that are valuable to us. But this function is perfectly well served even when you desire something that you believe you cannot acquire. For this reason, it is not irrational to desire X and, at the same time, to desire not-X; your desires, even in that case, can help you to see both what is worthwhile about X and what is worthwhile about not-X.

Intentions are not distinctive in virtue of what it feels like to have one. Intentions are a species of thought. And it is quite plausible to believe that thoughts, generally, require consciousness of some sort. If that is so, then intentions, too, require consciousness. However, there is no reason to think that intentions require any forms of consciousness that are not required, generally, for thought. That is, if it is possible, for instance, to have a belief or a desire or a whim in the absence of some form of consciousness, then it is possible, too, to have an intention in the absence of that form of consciousness.

One source of confusion about this is rooted in a failure to distinguish between intentions, which are mental states, and intentional actions, which are not.[7] It is plausible that human beings are good at knowing what they are doing, when they are acting intentionally, and at knowing that they are acting *intentionally* when they are. That is, we seem better at knowing these things than other things about ourselves—compare your knowledge of what you are doing with your

---

[7] See Chapter 21 by Parés-Pujolràs & Haggard on intentional actions.

knowledge of your blood type. But this does not imply that intentions are attended with some form of consciousness missing from some other forms of thought. The reason is that we might know what we're doing, or even that we're doing it intentionally, without deriving that knowledge from a form of consciousness that attends the relevant intention. Perhaps I know that I'm typing, for instance, in part thanks to the fact that I feel my fingers striking the keyboard. And perhaps I know that I am typing *intentionally* at least in part thanks to the fact that there's no other plausible explanation for why I'm typing. Or perhaps I know both these things without observation, including observation of any of my mental states, in something like the way in which I proprioceptively know the location of my limbs. It is only if we have a very narrow view of the possible sources of our special knowledge about our own intentional actions that we would reach the conclusion that we have some form of special access to our intentions that we lack with respect to other thoughts.

2

What is a will?

Pamela Hieronymi

This is a controversial question. We could start by saying that the will is the capacity for choice, or perhaps the capacity for voluntary or intentional action.[1] Doing so will simply spread the controversy to include "choice," "voluntary," or "intentional."

It is often thought that wills are the special possession of humans, persons, or morally responsible creatures. On this way of thinking, neither chickens nor chess-playing computers possess a will—even though, it might be admitted, chickens have desires on which they act and computers make choices. We might then try to update our working definition to say that the will is the capacity for *free* choice.[2] We would thereby increase the unclarity immensely. (Qualifying a difficult idea with "free" is an attempt to clarify by adding mud.)

To clean away some of this mud, I will sketch two broad, contrasting pictures.

On the first, I think more popular, picture, the will is a capacity to "step back" from, to somehow distance yourself from and reflect on, anything that might influence or determine your choices or actions—including your own natural tendencies, inclinations, or dispositions—and then to consciously determine, for yourself, undetermined by those influences, how you will choose or act.[3] On this picture, the will is a capacity to act "freely," where "free" action is action undetermined by anything other than the will of the person—including, importantly,

---

[1] See Chapter 1 by Yaffe on intention, Chapter 3 by Hieronymi on voluntary action, and Chapter 21 by Parés-Pujolràs & Haggard on intentional action.

[2] See Chapter 4 by Sinnott-Armstrong on freedom, Chapter 5 by O'Connor on free will, and Chapter 8 by Mele on free will.

[3] Do you "determine, for yourself, how you will choose" by making another choice? It is important to avoid a regress, at this point.

Pamela Hieronymi, *What is a will?* In: *Free Will.* Edited by: Uri Maoz and Walter Sinnott-Armstrong, Oxford University Press. © Oxford University Press 2022. DOI: 10.1093/oso/9780197572153.003.0002

the aspects of the person's psychology mentioned earlier—their tendencies, inclinations, dispositions. The will thus conceived is the ability to originate activity independently of any external influence.

So conceived, one might model the will as a kind of internal module from which originates some spontaneous or creative force. Such a module, one might further think, is the special possession of humans, persons, or morally responsible creatures.[4]

Those utilizing this kind of picture must navigate several hazards. First, the module must not itself become, or contain, a homunculus, lest they face a question about the will of that little person; the module is rather the special possession of a person, that which allows the person to act independently. Second, the picture must keep the outputs of the module independent of (completely independent of? only probabilistically determined by?) forces outside of it, including the person's broader psychology. But, finally, those advancing this picture must also, at the same time, address the question of how or why the output of this particular spontaneity-producing module is especially important or significant—why we should identify the *person's* activities with *that module's* activities. This last will be difficult: by design, the module operates independently of any external influence, but much (perhaps all) of what we identify as central to ourselves, as persons, lies outside of it. Thus, it seems the will must operate independently of many (perhaps all) of those psychological features or aspects that we usually identify as constituting the person, and yet its activities must be our own.

On the alternative picture, the will is not an independent module within the subject that originates a spontaneous or creative force. Rather, the will is that collection of more or less ordinary, interacting aspects of the person's psychology (their cares, concerns, beliefs, desires, commitments, fears, etc.) that generates intentional, voluntary, or responsible activity; it is the functioning together of those aspects of mind that account for human activity. On this picture, willed activity, or choice, is "free," not because it is specially independent or spontaneous but rather because it is distinctively "owned by," due to,

---

[4] See Kreiman's replies in Chapter 17 on the neural basis of the will.

or the responsibility of the individual whose activity it is—where the individual is identified not with a module for spontaneous or creative activity but as a complex psychological subject whose features are, ultimately, a product of nature and nurture.

Consider the following quotation from philosopher Harry Frankfurt (1988b, p. 84): "If we consider that a person's will is that by which he moves himself, then what he cares about is far more germane to the character of his will than the decisions or choices he makes." The quotation appears in a paper in which Frankfurt considers situations in which an individual cannot bring themselves either to make or to follow through with certain choices, because (perhaps to their own surprise) the choice is contrary to those things they care most deeply about. Such situations are not rare: you cannot betray a friend; you must accept a certain challenge or job opportunity. Frankfurt's paper thus highlights the potential shallowness and insignificance of mere, or "bare," choice or decision cut loose from the rest of the person.

In the quotation, Frankfurt *contrasts* the capacity for choice with the will. Again, he identifies the will as those aspects of the person by which the person moves themselves—by which the person is a self-mover. These include not only the person's cares and values but also their desires, convictions, commitments, beliefs, intentions, emotions, etc. These inform and can sometimes countermand our choices.[5] When they do countermand our choices, we are prevented, *by ourselves*, from making a choice: we find ourselves unwilling either to make it or to follow through on it.

Again, on this picture the will is not an independent module. It is, rather, that collection of more ordinary states of mind that generate the distinctive self-movement of the person. And that "self-movement" does not require independence or spontaneity. It must instead be identified in a different way.

This is the challenge faced by this second picture: Why is the interaction of these ordinary, influenceable, even manipulable states of mind especially significant? Why is *it* activity for which we are especially

---

[5] See Chapter 23 by Hopkins & Maoz on the neural correlates of beliefs and desires.

responsible? More, how can ordinary, influenced, manipulable states of mind generate activities free enough to ground responsibility?[6]

(Notice that, even though the second picture does not identify the will specifically with the capacity for choice, there may yet, on that picture, be such a capacity, and it may be something to investigate scientifically.[7] However, there is no need for that capacity to be especially spontaneous or independent of external influences.)

For my own part, I find the challenges faced by the first picture impossible to navigate simultaneously. The challenge faced by the second, in contrast, I believe can be met.

Follow-Up Questions

Mark Hallett

Is freedom a necessary characteristic of will? Also, in relation to the second picture, you write, "[W]illed activity, or choice, is 'free,' not because it is especially independent or spontaneous but rather because it is distinctively 'owned by,' due to, or the responsibility of the individual whose activity it is." If someone had a delusion or hallucination and acted on that basis, would the individual be responsible, would he own the act, and would it be free?

Walter Sinnott-Armstrong

You conclude that the first picture of the will that you discuss faces challenges that cannot be met, but the second picture can meet "the challenge" of identifying "the distinctive self-movement of the person" and which "ordinary states of mind" generate that self-movement. Please tell us more about *how* this challenge can be met. In particular, can neuroscience of the sort that Kreiman discusses[8] help defenders of the second picture meet this challenge? How or why not?

---

[6] See Chapter 14 by Yaffe on responsibility.
[7] See Chapter 17 by Kreiman on the neural basis of the will.
[8] See Chapter 17 by Kreiman on the neural basis of the will.

Liad Mudrik

When contrasting the two accounts, the author emphasized the independence of the will. But it was only the first account that referred to the person consciously deciding how to act. Is consciousness also required for the second definition of the will? If so, why is consciousness crucial for both definitions? If not, why is it included only in the first one?

Replies to Follow-Up Questions

Pamela Hieronymi

Response to Hallett

Is freedom a necessary characteristic of will? Thank you for this question, as it allows me to make an important correction and clarification. To answer: No, freedom is not a necessary characteristic of the will. Rather, the will is a capacity whose operation can be free or unfree, depending on whether that operation is hindered, constrained, defective, or interfered with. Or this is what I should say. However, sometimes "free" is used in a different way—and I am guilty of so using it earlier—to pick out not the absence of hinderance, constraint, etc., but rather to pick out self-determined movement, activities that in some way originate in or are especially owned by the person. In this second sense of "free," the will is meant to be the source of freedom, and so freedom *is* a necessary characteristic of its operation.

Thinking of the first conception of the will (as an internal module that originates creative or spontaneous force): its activity is self-owned, or counts as the movement of the person, simply *because* nothing external to it has influenced it. Thus, on the first conception, the two senses of "free" coincide. However, on the second picture of the will (as that collection of ordinary states that generate the person's self-movement), the activity of the will is self-owned because it is identified with the operating together of states of mind that we identify as the person (a person who is influenceable by things external to them). And that activity might be (not only influenced but) hindered, constrained, or interfered with: activity might be one's own and yet unfree. This brings us to Hallett's second question.

If someone has a delusion or a hallucination and acts on that basis, then the operation of their will has been hindered or is defective. Thus, the person is not free (in the first sense). They are also not responsible. (Not every hinderance or defect absolves one of responsibility, but these typically do.) Whether the action is the person's own, whether it counts as a self-determined act (or "free" in the second sense), depends on whether we include, within the person's will, the delusional states of mind. On the first picture of the will, we will not; the delusions and hallucinations will be external interferences with the person's will. However, on the second picture of the will, we *might* include the delusional states of mind among those that constitute the person's will. Whether or not we do so will depend on how fully the delusion is incorporated into the rest of the person's life, thought, or personality. Something that only occasionally shows up may be a foreign interference to the person's self-determination, while something that informs most of what the person thinks and does will instead be a feature of their (sadly, defective) will, something that is part of their self-determined activities. This would provide an example of activity that is self-determined (and so "free," in the second sense) and yet hindered or constrained (and so unfree, in the first sense).

Response to Sinnott-Armstrong
The challenge for the second picture of the will is to make clear why the interaction of ordinary, influenceable, even manipulable states of mind should count as the self-determination of the person, and so why *this* activity (and what follows from it) should be especially significant or why *it* grounds responsibility. I would argue that those aspects of our minds for which we can be asked for *our own* reasons—such as our beliefs, resentments, jealousies, pride, distrust—reveal what I would call our *take* on the world, our sense of what is good, important, worthwhile, horrible, unacceptable, disdainful, etc. Further, these states of mind, in being the sort of thing for which we have reasons, are, I would argue, "up to us" in a specific sense. Thus, we are, in this sense, self-determining. Because these are states of mind for which we have our

own reasons, we are also "answerable" for them—we can be asked for our reasons for them—and so we are, in this limited way, responsible for them. More must be said to explain why we are responsible for them, or for our actions, in any more robust sense.[9]

How does this way of responding to the challenge relate to brain areas? There are regions of the brain that seem *not* to contribute to things like beliefs or intentions or jealousies, and others that do—so, we have a start.[10] I would be cautious about assuming that states of mind we have learned to identify in our social interactions—states whose contours are delineated in part by their social functions (the belief that the butler did it, distrust of authority, an intention to quit smoking)—will enjoy a neat, one-to-one mapping to locations in the brain. But, surely, they will be realized in the brain in some way.

Response to Mudrik

Finally, Mudrik asks what is the role of consciousness in each picture of the will? I would suggest that consciousness seems important in the study of willed action because *something like* consciousness is crucial to action.[11] However, once we understand this crucial role, we will see that it is not important that the will *itself*, or that willing *itself*, be conscious.

We identify an event (the movement of a finger, say) as an *action* by identifying it as something that happened because someone *meant* for something to happen (they *meant* to move their finger, or to push the button, or maybe they meant to move their *other* finger but were confused by their visual input). To *mean* for something to happen is to have, in some very minimal sense, decided to bring it about.

It follows from the fact that every action requires someone meaning for something to happen that every action will involve some minimal sort of "having in mind"—some representation of what the actor meant to do. If we think of this having in mind as being conscious of, or aware of, what you mean to do, then every action will require *that*

---

[9] I say more about each of these issues in Hieronymi (2008, 2011, 2014, in progress).
[10] See Chapter 17 by Kreiman on the neural basis of the will and Chapter 23 by Hopkins & Maoz on the neural correlates of beliefs and desires.
[11] The answer that follows is also contained in Hieronymi (in progress).

form of awareness or consciousness. (If, instead, as I find more plausible, the required having in mind only *typically* involves awareness or consciousness, then it will be possible to act intentionally but subconsciously, without awareness.)

Importantly, though, the thing that you have in mind, in acting, is *what you mean to bring about:* the raising of your finger, say, or the pushing of the button. It is not awareness or consciousness of the psychological operations *by which* you will bring it about. And so, on neither account of the will do we yet have reason to expect ourselves to be conscious of our own will, itself, as it operates.

Why, then, is it so disconcerting to think that the "conscious will" is an "illusion" (as the Libet experiments were thought to show)? I submit this is because of our sense of what it is to *control* a thing.

We naturally think that we control a thing to the extent that we can bring it to be as we would have it to be—to the extent that we can, so to speak, conform it to what we have in mind. And so it seems that, to control a thing, we must have *it* in mind. We cannot control what we are unaware of, or what happens behind our backs. And so it can seem that, if our willing is *not* conscious, not something of which we are even aware, then it is not something we control. And if our willing, *itself*, is out of our control, then surely we are not self-determining.

The solution to this problem is not to ensure that we are aware of, or conscious of, our own willing. (Being aware of a thing does not, by itself, bring it under our control.) The solution is, instead, to recognize that our willing is special; it is not out of our own control just because we are not aware of it as we do it. (In fact, I would argue, our decision-making is up to us for the same reason and in the same way that our beliefs, our jealousies, our general take on the world is up to us.)

In sum, then, awareness, or consciousness, is important for control, and consciousness, or at least some sort of having in mind, is important for action. But it is not, I think, important that our willing, or our controlling activities themselves, be conscious.[12]

---

[12] See Chapter 13 by Bayne on consciousness and freedom of action or will.

3

When is an action voluntary?

Pamela Hieronymi

To borrow from Aristotle, "voluntary" is said in many ways. I here survey different contrasts one might want to mark with the words "voluntary" and "involuntary."

I start with a theory-driven usage that connects to my chapter in this volume asking "What is a will?"[1] In answer to that question I sketched two broad, contrasting pictures. On the first, the will is a capacity first to "step back" from all that would influence us and to determine for ourselves (perhaps "endogenously") how we will proceed. On this first picture, the will is a capacity for independent, or spontaneous, choice.

With this first conception of the will in hand, we could give a quick answer to the current question: Action is voluntary whenever it is the product of *this* special capacity, the will.

On the second conception of the will, the will is not a special capacity for independent or spontaneous choice; rather it is that collection of ordinary, influenceable, interacting aspects of the person's psychology that generate intentional, voluntary, or responsible activity.[2] On this picture, willed activity is "free" not because it is independent or spontaneous but rather because it is the unhindered self-movement of a person.[3]

On this second picture of the will, we identify a person's will as whatever it is that generates a distinctive or significant kind of activity— intentional or voluntary or responsible self-movement. Thus, we

---

[1] See Chapter 2 by Hieronymi on the will.
[2] See Chapter 21 by Parés-Pujolràs & Haggard on intentional action.
[3] See Chapter 4 by Sinnott-Armstrong on freedom, Chapter 5 by O'Connor on free will, and Chapter 8 by Mele on free will.

Pamela Hieronymi, *When is an action voluntary?* In: *Free Will*. Edited by: Uri Maoz and Walter Sinnott-Armstrong, Oxford University Press. © Oxford University Press 2022. DOI: 10.1093/oso/9780197572153.003.0003

cannot, on pain of circularity, identify voluntary action simply as the product of the person's will.

If we work in this second direction—from significant activity in, so to speak, rather than from special capacity out—the question of when an action is "voluntary" is much less straightforward. We must ask which human activities are intended by, up to, or the responsibility of the person. And these are not restatements of a single question but rather several interacting questions. Nonetheless, these are questions on which philosophical reflection has made some progress.

Philosopher Harry Frankfurt (1988c) identified *action* as purposive movement guided by an agent. Action thus contrasts with purposive movement guided by something *other than* an agent. And, indeed, this is one distinction we sometimes mark with the word "involuntary." Sometimes when we say that a movement is "involuntary," we mean that, like blinking, digestion, or homeostasis, it is a reflex or anatomical response. These are purposive movements, but they are not guided by the agent. Call this "involuntary$_1$" movement. It contrasts with what we can call "voluntary$_1$" movement—or, simply, "action."

Notice that, in making this first distinction, we need not start with a complete understanding of who or what an "agent" is, and work from the inside out. We can, instead, make progress by considering our intuitive, pre-theoretical understanding of the difference between, on the one hand, activities such as digesting and, on the other, actions such as cooking; we can then work from our understanding of such activities to an understanding of the agent. We might, for example, identify sufficiently flexible activities, those open to intelligent updating in novel situations, as those guided by the agent (as actions, voluntary$_1$ movements). We would thereby learn something about agents (they are capable of intelligent updating in light of new information).

Typically, in order to identify an event as a human action rather than a mere happening or an involuntary$_1$ process, we identify it as something that happened because the person *meant* for something to happen. Further, it is not important, in identifying an event as an action (a voluntary$_1$ movement), that it was spontaneous or that the person meant it independently of external influence. You might be commanded to act, or forced to act by your circumstances, or coerced. You nonetheless act; your behavior is not thereby rendered mere involuntary$_1$ movement. In my opinion, an important building block in scientific understanding is identifying the neurological differences between voluntary$_1$ and involuntary$_1$ movement—between those movements that are guided by the agent, those that happened because someone meant for something to happen, and those that are not.

Moving on: In identifying an action as something that happened because someone meant for *something* to happen, we must remember that what happened may not be what was meant. You meant to send the email to your mother, but instead you sent it to your supervisor. Sending the email to your supervisor was an action; in fact, it was your action. It was not an involuntary$_1$ movement. Yet we would not say that you sent the email to your supervisor voluntarily. Your action was not involuntary, yet also not voluntary. We need another distinction. Call this voluntary$_2$: An action is voluntary$_2$ if you meant to do *it*; voluntary$_2$ actions are done on purpose and successfully.

We could, at this point, add complication about unforeseen or foreseen but unintended consequences. Michael Bratman (1987), a philosopher of action, has offered a theory to map this terrain. We would then be studying what Bratman calls "planning agency," which one might think of as a kind of "executive capacity." Planning agency is the ability to intelligently, successfully bring about that which one (in some way) represents as to be brought about, the ability to form and execute intentions.

Again, if we were to study this, it would not be important that your plans are isolated from external influence, and thus it would not be important, in experimental design, to avoid commanded actions or actions done on cue. The fact that you were told to email your mother would not make your emailing her any more or less an exercise of your planning agency—no more or less voluntary$_2$.

However, we have not yet captured all we sometimes mean when classing an action as "voluntary." Aristotle (1999) provides what many think of as a paradigm instance of involuntary action: sailors throwing cargo overboard during a storm. The sailors throw the cargo overboard against their will: they are forced to do so by the storm. But throwing the cargo overboard is not an involuntary$_1$ movement. In fact, it is a voluntary$_2$ action: they set and successfully achieve their aim. Yet we want to say it was involuntary because they did it only under duress.

Thus, we need yet another distinction. Let us say that the sailor's action was involuntary$_3$. Cases of involuntary$_3$ action contrast with cases in which you are, we might say, happy with your choices. Duress is not the only source of unhappiness. We can add coercion, a sense of obligation, obedience, or need. In contrast, to act voluntarily$_3$ would be to act, so to speak, as a *volunteer*: to do what you do happily, without unwanted pressure—when you act, as we sometimes say, "of your own free will."

The problem encountered by the sailors, as well as by those subject to threats or coercion, is still *not* that they are subject to external influences or that they fail to act spontaneously. It is, rather, that all of their available options are contrary to their preferences. (If they were instead given an attractive offer for their cargo, they would then unload the cargo voluntarily, of their own free will—despite the fact that the offer provided a strong external influence.) Likewise, the problem with acting from obligation or need is that it does not align with your preferences. Lacking good options that align well with your preferences is a familiar and straightforward enough difficulty. But it is not a problem with your capacity for action; it is rather a problem with the circumstances in which that capacity must operate. It is, so to speak, a problem in life. It would, I think, be odd to think that the difference between acting as a volunteer, happily and without unwanted pressure, and acting under coercion, duress, obligation, command, etc., is to be studied by studying the human capacity for agency rather than by considering the circumstances in which such a capacity operates.

But we are not yet done. Not every case of failing to act "of your own free will" due to unhappy life circumstances is a case of failing to act as a volunteer: by use of deceit, indoctrination, propaganda, or clever advertising people can be *manipulated* into acting as a volunteer. The denizens of Huxley's (1932) *Brave New World* act voluntarily$_{1, 2,}$ and $_3$—they successfully form and execute plans, while acting happily, as volunteers—and yet they lack another important form of freedom. If one wants to capture this form of freedom with the word "voluntary," one will need yet another distinction: one might say that they are acting involuntarily$_4$.

Once again, the problem with involuntary$_4$ action is not that those who are manipulated, deceived, indoctrinated, etc. are subject to influence *per se*. Rather, the problem is that they are subject to a *problematic*

sort of influence, a kind of influence to which people ought not subject one another. It is famously hard to know how to draw the line between, say, education and indoctrination, coercion and persuasion, or
threats and offers. But drawing these lines will not be a matter of locating
a particular capacity in the individual that typically operates free of influence (or free of some degree of influence) and identifying when it is
influenced (or influenced to a greater degree). It will, instead, be a matter
of determining which *ways* of influencing others are permissible and
which are problematic. That is to say, drawing *these* lines will be a matter
of ethical reflection, not scientific or metaphysical discovery.

My suggestion, then, is that neuroscience should study (or, more
modestly, should *first* study) both the difference between voluntary$_1$
and involuntary$_1$ movement and our capacity for voluntary$_2$ action—
for successful planning agency. The forces that render our activities
involuntary$_3$ or involuntary$_4$ are problems encountered by humans
acting in a difficult world: they are problems in life, not problems with
our capacity for significant human action.

Follow-Up Questions

Liad Mudrik

When the author contrasts agent-guided purposive actions and
non-agent-guided purposive movements, like reflexes (following
Frankfurt), does the difference rest on the agent's conscious state? That
is, what determines whether an action is guided by an agent? Is it the
agent's consciousness of the purpose? If so, is it accurate to say that
consciousness of the purpose is the criterion for an action being voluntary? And if so, can any voluntary action be unconscious?

Walter Sinnott-Armstrong

How are the four kinds of voluntariness that you distinguish related to
moral responsibility? Are agents responsible for all of their voluntary
acts? Are agents not responsible for any of their involuntary acts?

Uri Maoz

You propose that neuroscience should focus on distinctions between in/voluntary$_1$ and in/voluntary$_2$. If this is because you think that more philosophical work needs to be done to better understand the distinctions between in/voluntary$_3$ and in/voluntary$_4$, that is one thing. But if you think that those latter distinctions are beyond neuroscience or that neuroscientific investigations are not useful there, please elaborate on your reasons for this.

In particular, one reason that we care about voluntariness in actions relates to moral responsibility. Supposedly agents whose acts are involuntary$_1$, involuntary$_2$, involuntary$_3$, or involuntary$_4$ are somehow less responsible (perhaps to different degrees?) than those who act voluntarily on all those accounts. So one reason that we might be interested in a neuroscientific account of the voluntariness of these actions is to possess a more objective account of an agent's responsibility, if any. To that extent, it appears that neuroscience can potentially help with the distinctions between voluntary and involuntary$_{1, 2, 3,}$ and $_4$. While it might be more straightforward to understand how to draw neuroscientific distinctions between in/voluntary$_1$ and in/voluntary$_2$, there is no reason to conclude that understanding whether a person is happy with their choice—as in the in/voluntary$_3$ distinction—is beyond neuroscience. For the in/voluntary$_4$ distinction the difference lies not within the agent but within the psychological or brain states of those influencing the agent. And those could be further studied by neuroscience. So neuroscience could also be brought to bear on in/voluntary$_4$. Do you agree?

Gabriel Kreiman

Imagine that you can measure every possible neuroscientific variable of interest: action potentials of every neuron, the concentration of every ion in every cell. You name the variables you want to measure, and you have them all! Is there any measurement that could tell you whether an action was voluntary or not? If you think that neuroscience is not sufficient, then feel free to add whatever variables you want, as long as they are really measurable things. You can add all the cells in the liver, the position of all the stars in the universe, or any kind of behavior, but it has to be a physically measurable variable and cannot be a "will" or an "intention," unless you tell me how to measure those things. Is there any empirical measurement that can tell us whether an action is or is not voluntary (in any of the four senses that you distinguish)?

Replies to Follow-Up Questions

Pamela Hieronymi

The distinctions I've drawn between different senses of "voluntary" concern the employment of a particular human capacity, the capacity for what might be called self-guided movement (voluntary$_1$ movement). The first distinction marks whether this capacity is in operation at all; the second marks the *skillful* and *successful* deployment of that capacity (voluntary$_2$); and the third and fourth mark whether the operation of that capacity is in some way hindered, constrained, or problematically interfered with or manipulated (voluntary$_3$ and $_4$).

Reply to Mudrik
The difference between self-guided (voluntary$_1$) movement, such as cooking or opening a door, and movement that is guided in some other way (involuntary$_1$ movement), such as digesting or blinking, is not whether either the movement or its guidance is something of which the agent is conscious. I am sometimes conscious of my digestion or of my blinking; the song going through my head is itself a conscious state. None is guided by me, in the relevant sense. On the other hand, I typically am not conscious of the ways in which my fingers move to grasp the doorknob, even as I grasp it purposively. Likewise for the ways I move my joints in order to catch the ball you threw. My conscious awareness of goings-on or of their purposes or of how they achieve their purpose does not correlate well with whether or not they are self-guided, voluntary$_1$.

Reply to Sinnott-Armstrong

Questions of responsibility overlap with these distinctions in voluntariness but do not map neatly onto them, because questions about responsibility concern not just the deployment of our capacity for self-guided movement but also the expectations or demands we can reasonably put on one another. I might be responsible for my digestion or my allergies because they are things I am rightly expected to manage. I am responsible for my short temper and my distrust, not only because they are things I can manage but also because they reveal my take on who can be trusted and what is of value—despite the fact that these are not voluntary. If I am acting under the influence of a powerful drug or suffering from a temporary delusion, then I might not be responsible for something I did voluntarily$_1$ and $_2$. Likewise, I might not be culpable for something I did voluntary$_{1, 2,}$ and $_3$, because I had been systematically deceived.

Reply to Maoz and Kreiman

The first and second senses of "voluntary" pick out a particularly important capacity of individual human beings: activity that is self-guided, and often skillfully and successfully so. Given a theory, we can figure out how to measure these. In the third sense, the agent either acts as a volunteer or else is in some way unhappy with their situation. Insofar as unhappiness is itself a psychological state, perhaps unhappiness interacts in interesting ways with the capacity for self-guided movement. To study this interaction, we would need some way to identify the two.

Some neurological studies adopt what seems to me a peculiar interpretation of "voluntary." They focus on activity that is "endogenous" in a very specific sense: the subject is not told exactly what to do or exactly when to do it but is instead asked to pick from a menu of options (left or right, this or that) at some unspecified point, after a certain point in time.[4] Granted, we are able to do this. We are able to resolve uncertainty, to do this-rather-than-that, now-rather-than-then, and to do so for no reason other than the need, or the desire, to move

---

[4] See Chapter 18 by Triggiani & Hallett on the neural processes of deciding what, when, and how to move.

on. This ability is an important component of our capacity for skillful, successful, self-guided movement; without it, we would not be able to execute many of our plans. But this particular ability seems to me just one part or aspect of our larger ability to execute our plans skillfully and successfully.

Further, I would argue that those moments at which we employ this "endogenous" ability are not ones in which our activity is especially "voluntary" or "free." The mere presence of "exogenous" factors which you might use to resolve the uncertainty (or which resolve the uncertainty for you, so to speak) does not itself show that your capacity for self-directed action has been in any way interfered with, hindered, or constrained. Nor does the presence of such factors itself show that what you do is any less self-directed or your own (unless you simply identify the self with this capacity for endogenous action, adopting something like what I called the first conception of the will in my chapter in this volume).[5] The presence of exogenous factors may in fact be part of your own complex plan. Whether or not they are, responding to them, or taking them to be reason-giving, may be something you do happily, as a volunteer.

The fourth sort of "voluntariness" (or, better, "freedom" or "liberty") concerns the ethical question of *which* ways of influencing others impinges on their freedom, understood as an ethical or political ideal—which ways of influencing others are illicit, ethically or politically. Some people think this ethical or political question should be answered by determining the conditions under which a person acts "freely" in some *other* sense; they would try to draw the line between offers and threats, or between persuasion and coercion, by first identifying the point at which the person's capacity for self-directed movement has been overpowered. They would suggest that the ethical line is crossed when that capacity is overpowered.

I think this is mistaken; we do not draw these lines by considering when someone's capacity has been overpowered. (When I hand over my money to save my life, I am acting voluntarily$_1$ and $_2$, and the fact that I am unhappy with the terms of our agreement, that I do not enter it voluntarily$_3$, does not show that you made a threat rather than an

---

[5] See Chapter 2 by Hieronymi on the will.

offer.) Rather, we first determine which ways of influencing others are unethical—which ways are disrespectful and so count as coercion or threats or manipulation rather than persuasion, offers, or influence—and then facts about whether the person was free, in this final sense, follow from this.

If I am right about the order of explanation, then, while impingements and constraints generated by ethically illicit behavior will, of course, be both constraints *on* and constraints *designed by* the operation of human brains, the relevant variables are unlikely to be neurological. An attempt to motivate the thought: the brain of a Beta in *Brave New World*, might, over some stretch of time, be neurologically indistinguishable from mine. Perhaps we have both, to this point, lived indistinguishable, unremarkable lives. Yet the Beta's freedom is greatly diminished due to her larger political situation: she has been systematically manipulated, while I have not. Of course, if we take a wide enough picture of space-time, we will find differences (there is "global supervenience" of the social on the physical), and some of them will be in human brains. But the fact that there will be such differences should not lead us to expect that the best way to study the phenomena we are after is by looking at those differences.[6]

---

[6] See Chapter 9 by Roskies on neuroscientific evidence and Chapter 10 by Bayne on behavioral experiments.

PART I, SECTION II

QUESTIONS ABOUT FREEDOM

4

What is freedom?

Walter Sinnott-Armstrong

Like many other words, "freedom" is best understood in terms of what it is *not*. To tell you that the mints on the hotel counter are free is to say that they cost *nothing*. To state that a seat is free is to say that it is *not* reserved. A country has free speech to the extent that its laws do *not* restrict what people say. In general, then, a claim that something is free is not positive but negative insofar as it claims that certain barriers, obstacles, or restrictions are *not* present. Different barriers are relevant in different cases, but freedom is always the absence of some kind of barrier.

The same goes for freedom of action. An agent is free to carry out an action just in case no relevant barrier prevents the agent from doing it. This answer is not supposed to settle any disputes, but it helps us understand disputes. When people argue about freedom of action, it is useful to ask, "Which kind of barrier is relevant?," "What are you denying?," or "Free as opposed to what?"

Incompatibilists claim that free action and will are incompatible with determinism.[1] Determinism is the claim that all events, including human actions and choices or wills, are determined by prior causes. Accordingly, incompatibilists are most concerned with barriers like causation or determination. They see causal determinism as incompatible with freedom of action and will because they hold that causation and determination make an agent unable to do or choose otherwise (Kane, 2007). They are right in this sense: if complete causal determinism holds, then no act is *free from causation* or from *determination*. That much should be obvious by definition.

---

[1] See Chapter 6 by O'Connor on determinism.

Walter Sinnott-Armstrong, *What is freedom?* In: *Free Will*. Edited by: Uri Maoz and Walter Sinnott-Armstrong, Oxford University Press. © Oxford University Press 2022. DOI: 10.1093/oso/9780197572153.003.0004

In contrast, compatibilists claim that free action and will are compatible with determinism. When they claim that an act is free, they are claiming only that it is *free from constraint* or from *excuse* in the sense that it is not constrained or excused. They are not denying that it is caused or determined, so they think that an act can be free even if it is caused or determined. On this compatibilist view, the agent acted freely if the agent was not constrained by being pushed or bound (force), threatened (coercion), misled or defrauded (mistake), driven by mental illness (compulsion or delusion), or anything like these excuses, which reduce moral responsibility.[2] These are the kinds of barriers that compatibilists see as relevant to freedom.

Compatibilists do not count all kinds of causation or determination as relevant barriers because, although some kinds of causes (such as pushing) do remove freedom, other kinds of causes do not excuse or constrain agents in the relevant way. In particular, agents who do something simply because they want to do it are not constrained from doing what they want, even if their desires cause them to do what they do. An agent's desires are not barriers that remove freedom of the kind that concerns compatibilists because desires are so different from force, coercion, mistake, or mental illness. The barriers that remove freedom in this view are only those external constraints that make agents unable to act as they want and that reduce moral responsibility.

The full story is longer and more complex, because compatibilists usually admit that some abnormal desires do remove freedom. For example, heroin addicts who struggle against their desires to take heroin still do what they want when they take heroin, but those acts are usually not free. Why not? Compatibilists disagree about this.

One answer (Frankfurt, 1988a) is that unwilling addicts' struggles show that they have a desire not to desire heroin, which is called a *second-order* desire because it is a desire about desires rather than about actions. When their first-order desire about actions (their desire to take heroin) conflicts with their second-order desire about desires (their desire not to desire to take heroin), their desires do not *mesh* or form a coherent whole. If they do not identify with their desire to take

---

[2] See Chapter 14 by Yaffe on responsibility.

heroin, then their drug use does not show what they are really like as a person. They are just sick temporarily, so their actions do not express their *deep self* (Sripada, 2016). An act is free on this view only when it is related in the right way to the agent's deep self, values, or second-order desires.

A different answer is that unwilling addicts take heroin even when they know that they have overriding reasons not to do so, such as when they know that police are watching or that they are likely to overdose or ruin their health. This pattern of emotion and behavior shows that these addicts are unable to stop using heroin, just as the fact that I would not stop a train from hitting me, even if I knew that I had an overriding reason to stop it, shows that I am unable to stop it. This *inability to respond to reasons* is a barrier that prevents these addicts from avoiding their harmful behaviors and thereby removes their freedom, according to these compatibilists (Gert & Duggan, 1979).[3]

Notice that such reasons-responsiveness and expression of one's deep self (or second-order desires) both come in degrees. Individual agents can respond to some reasons sometimes without responding to all reasons all the time, and desires can be more or less central to one's self or the kind of person one is. Hence, this compatibilist kind of freedom also comes in degrees. In contrast, the incompatiblist kind of freedom is either fully on or fully off if actions and choices are either fully determined or not fully determined.[4] Thus, these concepts of freedom are quite different in their logics and implications.

Indeed, these concepts are so different that we can have them both. One compromise position (Sinnott-Armstrong, 2012b) is that the incompatibilist kind of freedom (freedom from causation or determination) and the compatibilist kind of freedom (freedom from constraint or excuse) are both legitimate concepts of freedom. Each has a point because freedom from causation or determination is important to some issues (such as the place of humans in the natural world), but freedom from constraint or excuse is important to other issues (such as moral responsibility). When people debate furiously about freedom to act or to will, some people are talking about *freedom from causation*

---

[3] See Chapter 15 by Sinnott-Armstrong on reasons.
[4] See Chapter 7 by Hall & Vierkant on how free will can come in degrees.

or *determination*, whereas others are talking about *freedom from constraint* or *excuse*. Both are *correct* insofar as they are using their own concepts, but both are *incorrect* when they deny what their apparent opponents say because they mistakenly think that their opponents are using the same concept as they are. Their arguments misunderstand and miss each other.

On this view, then, we do not need to give up either concept, but we still do need to specify which of these concepts we are talking about. We also need to stop arguing. Maybe we can achieve peace by more charitably and accurately understanding what the other side means by "free."[5]

Follow-Up Questions

Yoni Amir and Liad Mudrik

Are freedom from causation and reasons-responsiveness incompatible? To explain, we assume that the more reasons-responsive (or even desire-responsive) a person is, the more predictable are her actions. And so, given that the reasons and desires are not arbitrary and should be similar if the situation is the same (hypothetically speaking, given some approximation), I am bound to act in the same way. Accordingly, is it accurate to say that the two concepts of free will (reasons-responsiveness and indeterminism) are in fact incompatible and that indeterminism can be applied only to arbitrary actions that have nothing to do with reasons?

Deniz Arıtürk

Why isn't freedom from causation or determination the kind of freedom required for moral responsibility? Is the distinction between causation and constraint merely a practical distinction that allows

---

[5] See Chapter 8 by Mele on free will.

us to *assign* responsibility to individuals in a potentially determined world, even if these people are not really responsible in a more fundamental way?

Antonio Ivano Triggiani and Mark Hallett

Thomas Hobbes (1651/1982) wrote Can God be included in the modern concept of freedom? If yes, how? And if we do not want to enter into a theological debate about the duality "God versus free will," can we consider the general concept of the "First Cause," as raised by Hobbes, useful in understanding freedom?

Eddy Nahmias

How can neuroscience help us determine which acts and agents are free from determination or free from constraint?

Replies to Follow-Up Questions

Walter Sinnott-Armstrong

These follow-up questions about my little answer raise many profound issues. I cannot address them all here, but I will discuss four.

Compatible?
Amir and Mudrik ask me whether reasons-responsiveness is incompatible with freedom from causation. I think so. To respond to a reason is to be caused by the facts that ground or constitute that reason.[6] My hunger is a reason for me to eat, and I respond to that reason when my

---

[6] See Chapter 15 by Sinnott-Armstrong on reasons.

hunger causes me to eat. Thus, reasons-responsiveness requires causation and is incompatible with freedom from causation.

I still might be free from determination if the reasons that cause me to act do not determine what I do, perhaps because the causal relation is probabilistic.[7] So reasons-responsiveness is compatible with freedom from determination.

But reasons-responsiveness is also compatible with determination. A reasons-responsive agent can be determined to respond "in the same way" whenever those reasons are present "if the situation is the same." Thus, reasons-responsiveness does not exclude determinism.

Even in cases of "arbitrary actions that have nothing to do with reasons," the agent can be caused and determined but still reasons-responsive. If I must choose between two identical boxes of the same cereal, I have no reason to pick one instead of the other, so there is no such reason that could cause my choice. Nonetheless, I can be caused to pick the left box, perhaps by a gust of wind or a group of neurons firing. And I can still be reasons-responsive if I am able to (and would) respond to reasons that favor one box over the other, if there were any such reasons.

Excuse?

Arıtürk asks me why causation and determination do not count as excuses that remove or reduce moral responsibility. This issue is normative, so it is difficult to answer without digging deep into moral theory. All I can do here is appeal to intuition (cf. Sinnott-Armstrong, 2016). Imagine that I promise to bring your dress to your wedding, and I remember my promise, but instead I go for a walk in the park, and your wedding is ruined. My desire for a walk together with my lack of concern for you cause me (and might even determine me) to take the walk, so I am not free from causation (or determination). Nonetheless, you would be angry with me, your anger would be justified, and I would be morally responsible. Why? Because this kind of cause—my desire and lack of concern—is not an excuse. Excuses

---

[7] See Chapter 5 by O'Connor on free will.

do include *some* kinds of causes, such as being pushed or coerced, but that does not show that *all* causes are excuses. Theories of moral responsibility aim to distinguish those causes that are excuses (and reduce moral responsibility) from those that are not (and do not). Wherever that line is drawn, if some causes are not excuses, then lack of freedom from causation does not imply lack of freedom from excuse.

Arıtürk suggests that this distinction between causes and excuses is "merely a practical distinction that allows us to *assign* responsibility to individuals." Yes, assigning responsibility is the goal. We need to distinguish those who are (and should be held) responsible from those who are not (and should not). That practical usefulness is a feature, not a bug, when the issue is normative, as moral responsibility is. If any supposedly "more fundamental" kind of responsibility is not normative, then it might not need to be useful practically, but it also might not seem so important.

God?

Triggiani and Hallett ask me how God and First Causes are related to freedom. That varies with the kind of freedom, again. If God and First Causes cause other events without being caused themselves, then they are free from causation. If humans are also free in this way, then humans are more like God, as some theologians claim. This might explain why we hold humans but not other animals morally responsible. These implications make some philosophers and theologians value freedom from causation. The other side of the coin, however, is that freedom from causation would undermine any physics that implies that every event is caused (with the possible exception of quantum events). To that extent, if humans were First Causes, they would not be part of the natural world that physics studies.

In contrast, humans do not have to be First Causes or God-like in order to be free from constraint or excuse. All they need is what Hobbes calls "absence of impediments to motion," where impediments are constraints. Freedom from constraint and from excuse are then compatible with physics and do not locate humans outside of nature.

Neuroscience?

Even if physics and other sciences do not exclude freedom, sciences still might not be useful in studying freedom.[8] Nahmias asks me whether neuroscience in particular can help us figure out which acts and agents are free. I do not see how, if the question is about freedom from causation or determination. Liljenström summarizes evidence for and against stochasticity in the brain and concludes that we do not know whether *any* brain events that affect our actions are stochastic or undetermined.[9] Even if we did know that *some* brain events are undetermined, neuroscience could not reveal *which* acts are stochastic or free from determination or causation.

In contrast, neuroscience could reveal that a particular agent is not free from constraint or excuse. If an agent recognizes overwhelming reasons not to commit some crime, but the agent commits the crime anyway, then neuroscience might provide evidence that this agent was unable to react to reasons because of a tumor (cf. Sinnott-Armstrong, 2012a) or neural dysfunction (as in addiction or compulsion). Neuroscience might also reveal dysfunctions that render an agent unable to correct a delusion or to remember a commitment and, hence, unable to recognize certain reasons.[10] In these ways, neuroscience might help to show which agents lack reasons-responsiveness and freedom from constraint or excuse.

---

[8] See Chapter 9 by Roskies on neuroscience and Chapter 10 by Bayne on behavioral experiments.

[9] See Chapter 29 by Liljenström on stochastic neural processes.

[10] See Chapter 17 by Kreiman on the neural basis of the will.

5

What is free will?

Timothy O'Connor

The term "free will" has ancient roots in philosophical discussions concerning the nature of human control over and moral responsibility for our actions. Its popular usage is of more recent vintage, doubtless influenced by increasing awareness of the scientific study of the human mind and behavior. The term "will" here signifies a common view that some human actions, and perhaps especially those crucial to grounding our moral responsibility, originate in "acts of will," i.e., choices.[1] Choice appears to us to be a power to settle on a course of action from among perceived alternatives. It is often preceded by deliberation that weighs reasons for competing actions, where reasons might be simple desires, intentions, or goals, working in tandem with corresponding beliefs about how such ends may be realized in the circumstances.[2]

What is it for the will, or choice, to be free? "Freedom" is a concept that has a negative component: speech is free in a political context when it is *not* subject to prosecution, and a consumer item is free if it is available without cost.[3] Similarly, acts of will or choices are free if they are not constrained by certain kinds of agent-internal or -external factors, e.g., uncontrollable desires or phobias that do not reflect the agent's considered view on what ought to be desired or avoided, or being subject to coercive psychological pressures (such as credible threats) that make it extremely difficult to choose in any other way. These examples suggest that free will may come in degrees, corresponding to the

---

[1] See Chapter 2 by Hieronymi on the will, Chapter 14 by Yaffe on responsibility, and Chapter 17 by Kreiman on the neural basis of the will.
[2] See Chapter 15 by Sinnott-Armstrong on reasons and Chapter 19 by Bold et al. on deliberate decisions.
[3] See Chapter 4 by Sinnott-Armstrong on freedom.

Timothy O'Connor, *What is free will?* In: *Free Will*. Edited by: Uri Maoz and Walter Sinnott-Armstrong, Oxford University Press. © Oxford University Press 2022. DOI: 10.1093/oso/9780197572153.003.0005

intensity of the phobia or external pressure, rather than being an "all or nothing" ability.[4]

Some philosophers maintain that if you choose freely, no set of prior factors—whether perceived as coercive or not—necessitated your having so chosen. Consistent with all the actual causal influences, you might have chosen otherwise or continued to actively deliberate or passively remained unsettled. They go on to argue that if agent-internal factors (neural or psychological) causally determine that the agent will make a particular choice at a given time and place, they do thereby necessitate her choosing as she does, and so the agent does not choose freely. (Their reasoning in a nutshell: I cannot be free with respect to any choice that is *a strict consequence of factors over which I have no control*. I now lack control over the internal state and the external circumstance I am in, as well as the causal laws governing their dynamic evolution. If the laws are deterministic, then each of my subsequent choices will be a strict consequence of such factors, and so unfree.)[5] Hence, if I am to choose freely, the relevant causal laws must be probabilistic only. With probabilistic laws, the internal determinants of my choices might narrow my alternatives and make the remaining options more or less probable to some degree. But that would still leave open which option I decide upon from among those alternatives. It bears emphasis that most philosophers who deem freedom to be incompatible with determinism do not deem it incompatible with causation; causes influence the direction of an outcome, and influences need not necessitate a precise outcome.[6] It is sometimes objected that if choices had merely probabilistic causal influences, they would be "random," occurring partly by chance. Such an objection likely depends on a reductionistic assumption that the only way that choices could be caused probabilistically would be "bottom up," through an amplification of quantum indeterminacies within the brain.[7] But free will, and especially paradigm instances involving conscious deliberation, is more naturally thought of as a kind of system-level, "top-down" exercise of

---

[4] See Chapter 7 by Hall & Vierkant on free will in degrees.
[5] See Chapter 6 by O'Connor on determinism.
[6] For detailed discussion of this distinction with reference to contemporary scientific examples, see Hitchcock (2018).
[7] See Chapter 29 by Liljenström on stochastic neural processes.

control. It is not merely the absence of certain constraints; it is a power to determine what conditioning factors leave unsettled.[8]

Other philosophers ("compatibilists") contend that the sense in which causes in a deterministic universe necessitate an agent's choosing as she did is consistent with her choosing freely. As they see it, choosing freely is simply to choose in accordance with one's preferences, free of abnormal distorting influences such as phobias, brain tumors, or coercive pressures. It is *constituted* by the regular efficacy of one's preferences in producing actions intended to satisfy them. If this more modest understanding of free will is correct, the only way that science could credibly challenge the reality of free will would be to give evidence that human choices are *typically* determined by factors *other than* the very desires, aims, and so on that make the choice seem attractive to the agent at the level of conscious awareness. Some scientists argue that there is evidence of such unwitting and non-rational determination of human choices (e.g., Wegner, 2002), but the gap between agents' occasionally being deceived and regularly being deceived about their own motivations is not closed (compare occasional sense-perceptual illusion vs. systematically faulty perception). Furthermore, a *science*-based argument for systematic motivational deception may be self-undercutting, since scientific practice arguably presupposes that scientists are *not* regularly and badly deceived about what they are doing.

Simple reflection suggests and recent psychology confirms that much of what we do is automated, carried out without focal awareness or choice of any kind. If moral responsibility requires that we are able to freely choose among alternatives, does it follow that we are responsible only for a small subset of our actions—those that do result from choice? No, we might be responsible for automated actions provided that they either trace in some way to past, character-forming free choices we've made or are consonant with the attitudes that we consciously affirm (and could have failed to affirm, given different past choices), even if we never explicitly deliberated about them. There

---

[8] For detailed analysis of the relationship of physical and mental causation and the reductionism/emergence debate, see Robb and Heil (2019) and O'Connor (2020), respectively.

is a degreed, *historical* dimension to free will, which requires that the unfolding process—partly conscious, much unconscious; partly chosen, much unchosen—whereby we become who we are is psychologically "integrated," constituting a reasonably coherent unfolding narrative (see Mele, 1995, pp. 131–176).[9]

Follow-Up Questions

Yoni Amir and Liad Mudrik

If freedom is really a matter of degree, is there some threshold for determining how much freedom is enough to deem an action free, or should we change the question from "Is this action free?" to "How much freedom is involved in this action?"

Deniz Arıtürk

What distinguishes abnormal from normal distorting influences? Are certain influences more freedom-diminishing, such as ones that are identifiable in a brain scan, appear later in life, or afflict a smaller portion of the population? If so, why? If not, how else can we distinguish abnormal influences from normal influences on behavior?

Tierra Lynch

Many factors (aside from more concrete external forces such as coercion) seem to erode free will. For example, forces such as advertising, manipulation, etc. can steer our actions in ways that make us deviate from our otherwise free decisions. These forces saturate our daily lives. Are any of our choices free from external influence? Would it be important to quantify the extent of external influence and grade the

---

[9] For comprehensive discussion of issues that the concept of free will raises and the ways that philosophers have theorized about them, see O'Connor and Franklin (2020).

blameworthiness accordingly? Additionally, circumstances like adverse childhood events, social determinants, trauma, etc. can create environments with choices so limited that it seems unreasonable to expect a normal rational person to have the "strength of will" to do otherwise. Such experiences and extenuating circumstances also seem to modulate our assignment of blame. In a society where factors like these inform the decision-making of a vast proportion of its people, how would this affect the practical usability of this system of moral responsibility? How could we ever assign moral responsibility in practice and not just theory?

Eddy Nahmias

What sort of evidence is relevant to sorting out whether probabilistic top-down causation actually exists?

Replies to Follow-Up Questions

Timothy O'Connor

Free will is easier to identify in an abstract and idealized thought-scenario than in the wild, owing to the patchwork and fragile reality of human psychology.

Amir and Mudrik ask whether we should dispense with the question of whether we have free will or not and instead try to develop measures for quantifying its degree in particular cases. I propose a slightly more complicated analysis: a set of baseline conditions are necessary and sufficient for deciding and acting freely (for having free will at all), *and* it comes in degrees of "intensity" and scope. The baseline conditions are being a conscious agent who is aware of a plurality of options,[10] chooses in a reasons-responsive manner,[11] and (more

---

[10] See Chapter 12 by Block, Chapter 13 by Bayne, and Chapter 24 by Mudrik & Schurger, all on consciousness.

[11] See Chapter 4 by Sinnott-Armstrong on freedom and Chapter 15 by Sinnott-Armstrong on reasons.

controversially) whose so choosing is not causally determined.[12] The degreed dimensions of freedom concern one's conscious awareness of influences and options. To what extent is the agent aware of (i) rational or nonrational factors significantly influencing his choice, (ii) valuable options readily available to him, and (iii) objective reasons for (dis) favoring those options?

It is because free will is strongly associated with moral responsibility that all three variable factors are relevant. An agent who is self-determining and highly self-aware but is more or less deluded about how his circumstance aligns with goals he has or ought to have is not responsible for failing to pursue a better course of which he is ignorant—unless he is *culpably* ignorant, owing to past choices or failures to act that have conditioned him to see things in a highly distorted way. Lynch and Arıtürk rightly observe that, when we consider the variety of influences on our behavior and our variable awareness of them, the category of "choice-distorting" (and so freedom-diminishing) influences cannot be sharply defined and their impact cannot be easily measured. This messiness is a consequence of two basic realities concerning the human mind that have perhaps long been dimly grasped but which scientific study forefronts. First, our psychology is highly fragile, profoundly affected by passing bouts of sleep-deprivation and mood swings and by the onset of disease and other protracted physiological conditions, as well as being susceptible to certain kinds of subtle psychological manipulation. With respect to the latter, when assessing responsibility, we must make rough and ready assumptions about people's ability to cultivate a critical mind, inform themselves concerning their own vulnerability, and take reasonable steps to avoid or counteract such manipulation. But we also must recognize that extreme forms of social conditioning or psychological trauma, especially early in life, cannot be easily curtailed, not least because those subject to them often don't *want* to see their world differently. What is common to freedom-eroding influences in contrast with everyday persuasion is that freedom-eroding influences distort a person's perception of her

---

[12] See Chapter 6 by O'Connor on determinism.

environment or the value of alternatives before her, or they bypass her capacity for critical reflection.

The second pertinent complicating reality is that we are curiously hybrid creatures, governed by a patchwork of mammalian machinery and drives that interact with our (more recently evolved) distinctively personal capacities and attitudes in an unstable dynamism over which we have limited control. Nahmias asks what evidence there is that we *ever* exert probabilistic "top-down" control over behavior—perhaps it's mechanistic forces through and through. There are two independent constituent features at issue here. I set aside whether human action unfolds probabilistically, as I address this in the replies to the determinism chapter. Concerning the reality of "top-down" causation,[13] the short answer is that we necessarily—and reasonably—*presuppose* it. All rational inquiry presupposes that communities of inquirers can knowingly and effectively investigate their world, responsive to norms of evidence and theorizing, and such a presupposition is inconsistent, I believe, with a thoroughgoing reductionism.[14] The scientist, of course, prefers to have independent evidence for such a presupposition. It is possible in principle but hard to envision in practice because of practical measurement and computational limitations.[15] When we consider the behavior of any complex system, no matter how strikingly distinctive, there is no inconsistency in the reductionist hypothesis that the behavior is entirely fixed by the same physical dynamics at play in every other context. If we were a LaPlacian intelligence (who knows all about nature's forces and particles), we could apply particle physics to all the elements of a living person (without disturbing brain function!), run the computations for each of their expected trajectories over a significant interval, and determine whether or not the predicted outcome is consistent with observed results. Alas, we are not so marvelous, so the best we might hope for would be to demonstrate the success or failure of certain explanatory reductions, under necessary simplifying

---

[13] See Chapter 20 by Hallett on higher-level control and Chapter 25 by Gavenas et al. on conscious control.

[14] See replies in Chapter 26 by Hopkins et al. on consensus on consciousness and volition.

[15] See Chapter 9 by Roskies on neuroscientific evidence, Chapter 27 by Hallett & Schurger on neuroimaging, and Chapter 30 by Kreiman et al. on computational models.

assumptions, within localized areas in a stepwise manner (neural activity to constituting chemical activity, and psychologically described activity to neural activity) and also (crucially) at the system's boundary conditions. If attempted reductions fail, one also has to reckon with the persistent possibility that we have simply gotten the underlying theory wrong in some respects. At present, the competing reductionist and "strong emergentist" hypotheses in relation to most *any* interesting complex systems have the status of conjectures—unrefuted, but also not significantly confirmed. But we are agents who act as embedded parts of the cosmos rather than being disembodied LaPlacian passive observers. Consequently, we all of us inevitably (and reasonably!) lay strong bets in our practice as to which of the competing visions will be best supported in the long run by the steady accumulation of empirical evidence.

6

Can there be free will in a
determined universe?

Timothy O'Connor

Some philosophers contend that free will is compatible with *causal determinism,* the thesis that our universe evolves in such a way that an ideally complete physical description of it at any given time and the true laws of nature together entail unique total universe states for all other times. (The true laws are what physics, biology, etc. are trying to figure out, not necessarily those that figure in their best current theories.)[1] If causal determinism were false and the laws were fundamentally statistical instead, only a set of weighted possibilities would be entailed. Deterministic laws, you might say, describe transitions in a domain as one-to-one rather than one-to-many. "Compatibilists" about free will and determinism motivate their position in different ways, but a guiding core proposal is that to act with free will is to do what you most want to do, absent any unusual internal or external obstacles to doing so (e.g., phobias, compulsions, other cognitive disorders, or strong coercion). Determinism *per se* does not undermine free will; only being determined by the wrong sorts of causes does.[2]

Philosopher-psychologist William James (d. 1910) judged such compatibilism to be a "quagmire of evasion." As he and other philosophers see it, a necessary condition on free will is that *the future is not fully settled* but instead is open to alternative possibilities. Unlike the fully determinate past, it is a space of branching possibilities, and free choices fix which branch at a forking node

---

[1] See Chapter 26 by Hopkins et al. on consensus on consciousness and volition.
[2] See also Chapter 4 by Sinnott-Armstrong on freedom, Chapter 5 by O'Connor on free will, and Chapter 8 by Mele on free will.

Timothy O'Connor, *Can there be free will in a determined universe?* In: *Free Will.* Edited by: Uri Maoz and Walter Sinnott-Armstrong, Oxford University Press. © Oxford University Press 2022. DOI: 10.1093/oso/9780197572153.003.0006

represents the actual future up to that juncture. A long-standing argument for *incompatibilism* between free will and determinism goes as follows: What the distant past was like and what the true laws of nature are are both entirely beyond our control. They are *already* settled. If our universe unfolds deterministically (remember: one-to-one transitions), those two settled matters yield a true and complete description of every future state, including every ostensibly free human choice. But, in general, it is not open to us to "settle" whether or not a future event occurs when it *deductively follows from* already settled truths. So if determinism is true, it is not now up to us what our future choices will be: we have no ability to make a choice other than the one that was settled long ago. We are not *free*.

In contrast, compatibilists maintain that in the relevant sense of "able," we *are* able to do what we are determined not to do. A tractor is able to run in reverse even while it is determined to run forward under the farmer's guidance. Its running in reverse requires only a different input, such as the farmer's shifting the gear. Likewise, I am able to stand even while my sitting is determined by factors such as my desire to begin drafting this chapter. My standing requires only some difference of *psychological* input, such as my feeling an urge to stretch. In general, our exercising an ability to do something we are causally determined not to do requires that something about the past or the natural laws would have been different. Now we of course are in no position to change the past or the laws. But that is irrelevant (says the compatibilist), since having an ability does not require that all the conditions on exercising the ability be present, or that we are able to arrange for them to be so.

Incompatibilists respond that this way of reconciling freedom and determinism illicitly switches the focus from what I am *here and now* able to do, given how things actually are, to what I have a general ability to do, even if I am determined to do otherwise. I retain my *general ability* to play basketball even if you have bound me to a chair, rendering me *unable-here-and-now* to exercise that ability. Free will requires our being able here and now to do otherwise, given how things actually stand, and not merely to possess a general unexercised ability.

Countermoves abound. Here one further issue warrants comment, as it strongly influences how philosophers come down on our question. It concerns the nature of causal laws and of causal "necessity." Incompatibilists often think of causal laws as fundamental features of the world that *drive* its forward evolution, saying things such as "It is the *nature* of this ion channel to open or close in response to certain changes in the voltage gradient." Compatibilists, by contrast, tend to think of causal laws, even basic physical laws, as derivative features, mere true generalizations over the distribution of independent local events within the space-time manifold. A true description of the deep, regular patterns running through them are what we call "physical laws." There is no deep necessity to events being arrayed as they are. Had the pattern differed, the laws would have differed. The contingent distribution of events, then, determine what the laws/generalizations are, and not vice versa, much as individual pixels of paint determine the overall character of a painting. If laws of nature merely record deep generalizations of a certain kind, then the way that laws and facts about the past together "determine" the future can easily seem irrelevant, in that they are not prior realities "making" the future to be what it is.[3]

Incompatibilists such as James do not hold that the threat of determinism would automatically go away if basic physics proves to be indeterministic. For it is consistent with this conjecture that reality be "nearly" deterministic *at the macroscopic level*, at least when it comes to natural macroscopic structures such as the human brain. (Small-scale indeterminacies could effectively "cancel out," leaving the macroscopically defined outcome 99.99 . . . % probable.) If incompatibilism is correct, then for us to have free will, it's plausibly not enough that our actions are not strictly determined with quantum-scale precision: there needs to have been *non-negligible* chances that some of our *choices* have gone differently. (This naturally raises the question of how significant a chance of alternative choices being made freedom requires. But we cannot explore that here.) Would natural mechanisms of "amplification" of highly localized, small-scale indeterminacies suffice? A few think so (e.g., Kane, 1996), but most agree that free will is a paradigmatic system-level or "top-down" form of control. If we are free, it must be that we ourselves determine which choices we make.

---

[3] For further discussion of the two competing visions concerning the nature of natural laws, see Carroll (2016).

How to understand in empirical terms the interplay of bottom-up and top-down features of complex systems such as ourselves is among the most vexed questions in the philosophy of nature.[4]

Given that the status of determinism in physics and in neuroscience are open questions,[5] and supposing that incompatibilism is correct, does it follow that we do not currently know whether we have free will?[6] To answer this question, we should start by reflecting on what kinds of general claims we can know (or reasonably believe) about the empirical world prior to scientific investigation—beliefs we perhaps necessarily *bring* to such investigation. Plausible examples include belief in the rough-and-ready reliability of human cognitive capacities and that the world has a regular causal order. Might our limited freedom, too, be among those claims we are entitled to *start from* rather than reason *to* via empirical evidence?[7]

Follow-Up Questions

Gabriel Kreiman

It seems that it is possible to come up with reasonable "excuses" for a fully deterministic world where free will still mysteriously survives, along the lines articulated in this nice response. In this case, is there any conceivable empirical observation that would ever show that there is no such thing as free will? In an oversimplified fashion, I think of scientific theories as statements that can be proven wrong (according to Popper, 1934/2005). If the answer to this follow-up question is no, then should we say that free will is *not* a scientific theory? If free will is not a scientific theory, how is it not akin to "Black holes contain invisible smiling dwarfs that have no mass and do not obey the laws of physics"?

---

[4] For detailed analysis of the reductionism/emergence debate, with reference to going scientific theories and models at multiple levels of the physical hierarchy, see O'Connor (2020).

[5] See Chapter 29 by Liljenström on stochastic neural processes.

[6] See Chapter 8 by Mele on free will.

[7] This question is explored in O'Connor (2019).

Lucas Jeay-Bizot

You write that "compatibilists maintain that in the relevant sense of 'able' we *are* able to do what we are determined not to do. . . . our exercising an ability to do something we are causally determined not to do requires that something about the past or the natural laws would have been different." This seems to assume that this something *could* have been different. How would the compatibilist justify that things could have been different in a determined universe?

You also write that, according to compatibilists, "[d]eterminism *per se* does not undermine free will; only being determined by the wrong sorts of causes does." The *right* sorts of causes are psychological states constituting "what you most want to do." Is it a matter of chance whether your action on a given occasion has the right vs. the wrong sort of causes? If so, does that not call into question your free will? If not, how is this consistent with the compatibilist claim that things could have been different in a determined universe?

Walter Sinnott-Armstrong

Your closing paragraph asks, "Might our limited freedom, too, be among those claims we are entitled to *start from* rather than reason *to* via empirical evidence?" However, if we start by assuming freedom, and if freedom is incompatible with determinism, then we assume from the very beginning that determinism is false. On this assumption, we do not need to consider any of the scientific research that Liljenström surveys.[8] Liljenström concludes, "The bottom line is that both indeterministic and deterministic processes can account for the observed (unpredictable) fluctuations in neural processes." This conclusion suggests that we do not know (yet?) whether our actions are or are not determined.[9] In contrast, you seem to suggest that we "might" know that some of our actions are not determined. Is it really that easy to settle the complex questions that Liljenström discusses?

---

[8] See Chapter 29 by Liljenström on stochastic neural processes.
[9] See Chapter 9 by Roskies on neuroscientific evidence about free will.

Replies to Follow-Up Questions

Timothy O'Connor

Two of these follow-up questions express skepticism regarding free will–determinism compatibilism. Kreiman worries that it may render free will an unfalsifiable thesis, on a par with outlandish hypotheses concerning invisible realities that operate in inaccessible regions of the universe. The compatibilist will say, to the contrary, that free will as she understands it is a readily observed and confirmable phenomenon: most mature human beings appear to regularly act in ways that they plausibly take to contribute to attaining what they want. The more likely charge here is that the existence of human free will on this conception is boringly obvious, and so of no scientific interest. And then one begins to suspect that it deflates the ordinary conception. Isn't it widely regarded as a distinctively human capacity that contributes to the dignity of persons? Some compatibilists do take the existence of free will to be obvious, believing that controversy concerning its reality is manufactured, a product of confusion regarding what it means to will or choose "freely." Others, however, allow that it entails commitments that could conceivably be falsified by mature psychological science (even while denying that causal determinism is one of those commitments). On their view, freedom of will requires a complex psychology of a particular kind, including sensitivity to objective norms and values and the ability to prioritize these over mere desires.

Jeay-Bizot doubts that the compatibilist's joint commitment to determinism and to the claim that agents' choices often "could have been different" is coherent. Determinism entails that there are sufficient causes for every such choice, causes that *preclude* alternatives. In reply, compatibilists will say that we need to distinguish an event's being physically determined from its being *absolutely necessary*, on a par with truths of pure mathematics. The laws describing our world's unfolding are unchanging, but they are also just one way a world might be structured, as is reflected in the fact that we need to do experiments, not mathematical theorem-proving, to establish what they are. A determined agent's having been able to choose otherwise[10] than she did

---

[10] See Chapter 8 by Mele on free will.

amounts to the fact that her ability to choose is *responsive* to reasons.[11] Sensitive to her constantly changing environment, her choices vary accordingly. Human technology is rife with devices that are determined on any occasion to yield a particular result (a computer's analyzing a data set, say) but could have carried out a different function (computing the programmer's taxes). Computers do not (yet?) have free will, but they illustrate the point of behavioral flexibility: having abilities that outstrip the one behavior that their causal inputs at any particular time determine them to do.[12]

Jeay-Bizot further notes that if, as compatibilists claim, choosing freely requires being determined by the "right sort" of causes (roughly, those that do not bypass the capacity for practical deliberation or distort its proper functioning), then, since the agent doesn't control whether her causes are of the right sort, her choosing freely doesn't seem to be an *achievement*. As far as she is concerned, it's a matter of luck. Here, the compatibilist is apt to bite the bullet: for the most part, whether or not you regularly act with free will is not up to you. Like most basic capacities, you've either been dealt that hand or you have not, although thankfully in this case it is widely if not quite universally shared. However, also as with other capacities, you have the ability to act in ways that affect the degree of freedom exhibited in future choices, positively or negatively. (You cannot guarantee robust future freedom, but you can guarantee its absence, so choose carefully!) Whether this means that, in the final analysis, free acts are not in some deep sense a credit to the agent is an excellent question that we leave the reader to ponder. (The incompatibilist sees this as a vulnerability in his opponent's position that makes attractive a more demanding understanding of free will that is worth wanting.)

Finally, Sinnott-Armstrong sees an odd implication of the opposing incompatibilist position (held by this author) on our question. If (as many assume) we know prior to scientific investigation that we are free, and freedom entails that our choices are undetermined, then it seems to follow that we can know from our armchairs that our actions are undetermined. But this question is currently being investigated

---

[11] See Chapter 15 by Sinnott-Armstrong on reasons.
[12] See Chapter 11 by Hall & Vierkant on artificial intelligence.

and is regarded as difficult to resolve by knowledgeable theorists! Many scientists will judge it absurd that we could know in this way the answer to a complex empirical truth. But as we noted at the end of our original answer, there are empirical theses about the reliability of our own cognitive capacities and the causal regularity of physical reality that we must assume in order to begin to do science. If these starting points to inquiry are not items that we know, then how can the fruits of such inquiry be known, either? Perhaps belief in our freedom is a proper starting point for our engagement in and investigation of *social* reality. But we might also suppose it to be a *defeasible* starting point, one that evidence could lead us to abandon. (It seems scarcely imaginable that we would altogether abandon moral practice shaped by this assumption, but perhaps it could undergo significant revision in the light of future discovery, such that we continue to think of ourselves as free and responsible, but only in attenuated senses.) In any case, the pleasing picture of scientists disinterestedly pursuing the truth is false to history, especially in matters of profound human concern. And that's okay. For history also shows that, while difficult, it is possible to have commitments on matters that are not scientifically settled while being willing to examine the full range of evidence. Let the investigations of Liljenström and colleagues continue, even as some of us place our large bets.

7

Does free will come in degrees?

Jonathan Hall and Tillmann Vierkant

It is easy to get the impression that the question of whether one has
free will is binary, so that one therefore either has it or doesn't have it.
This is because it is intuitive to think that free will requires the ability
to do otherwise (in a specific sense),[1] and either we have this ability
or we don't. If the macro-world is deterministic, then the past fully
determines the future, and there are no alternative possibilities. If
free will requires the ability to do otherwise and determinism is true,
then there is no free will, not even a little bit. This is what so-called
incompatibilists think.

Compatibilists, however, argue that an agent can have free will
without having the ability to do otherwise, as understood by the
incompatibilists. In order to answer whether free will might come
in degrees (or "is continuous"), we will look at four components,
highlighted in italics, that compatibilists have considered when
exploring whether an agent has free will:

Freedom to *act in line with intentions*
that one formed *consciously*,
through the exercise of normally functioning *rational machinery*
and with *no external interference*.

Some philosophers will feel that only a subset of these four are necessary for an agent to have free will, while others might demand additional or different components.[2] Although this list is not supposed to

---

[1] See Chapter 8 by Mele on free will.
[2] See Chapter 4 by Sinnott-Armstrong on freedom, Chapter 5 by O'Connor on free
will, and Chapter 8 by Mele on free will.

Jonathan Hall and Tillmann Vierkant, *Does free will come in degrees?* In: *Free Will*. Edited by: Uri
Maoz and Walter Sinnott-Armstrong, Oxford University Press. © Oxford University Press 2022.
DOI: 10.1093/oso/9780197572153.003.0007

be exhaustive, hopefully it will convey the idea of how it might not be implausible to have a continuous measure of free will. We will proceed in a way that allows readers to pick and mix components depending on their views.

For many (but not all) philosophers, some form of *conscious aware- ness* of intentions and/or of the formation of intentions is necessary for free will, because otherwise one might worry that the rational mechanisms that are available in our cognitive machinery might by- pass the agent (see, e.g., Vargas, 2013).[3] For example, a sleepwalking agent is usually not considered to freely form intentions or make choices, despite the fact that their behavior can look surprisingly ra- tional. However, a driver consciously focused on a podcast can in- tentionally slow down at a traffic light, despite not being consciously aware of any intention to slow down. More generally, consciousness of intention and choice becomes less and less necessary as actions are re- peated. If it is right to think of conscious control as coming in degrees, and if consciousness is necessary for freedom, then freedom might also come in degrees.

When considering an agent's *rational machinery*, the words "nor- mally functioning" are appropriate because no agent is capable of ideal rationality in all situations. Humans' deliberative abilities are ca- pacity constrained, and it requires skill and effort to overcome these constraints. These capacities need to develop, so that children will have less rational machinery at their disposal than will fully grown adults. Exhaustion may also reduce rational functioning by degrees. In more extreme cases, a head injury, for example, can change an agent's per- sonality, reducing her ability to control her anger. Intuitively, these cases involve a lower degree of free will than in fully functioning agents.

External interference can also come in degrees, ranging from full judgment override by an evil demon to an irrelevant short-term dis- traction, such as by a surprising noise or flash. It can include attempts to hack an agent's rational machinery or to bias the available evidence, such as by misleading advertisements. The agent can be cognizant or

---

[3] See Chapter 13 by Bayne on consciousness and freedom.

entirely unaware of the attempted interference. As a function of these degrees, external interference will be easier or harder to resist.

No human is ever entirely unconstrained in their ability *to act in line with their intentions*. Humans, like all natural creatures, have physical and mental limitations. However, many constraints can be overcome through training, effort, or planning. An addict who forms an intention to quit may succumb to temptation, but that doesn't necessarily imply that the temptation is irresistible. Resistance can be possible through effortful willpower or distraction or by structuring the environment in a helpful way (so-called tying-to-the-mast strategies, alluding to Ulysses's encounter with the sirens; see, e.g., Vierkant, 2015). Although addicts may feel a "loss" of agency, this should be interpreted as a partial rather than complete loss of agency and freedom.

Before we end, two points should quickly be added. We started off by saying that incompatibilists might think that freedom of the will is binary because whether we have the ability to do otherwise or not is binary. However, even if that is true, it does not mean that incompatibilists cannot also think that freedom of the will comes in degrees. No incompatibilist thinks that the ability to do otherwise is the only necessary condition for free will, and most will also use some or all of the compatibilist components we have discussed. As these do come in degrees, so might freedom of the will for these incompatibilists.

Finally, despite the fact that all of the discussed components do come in degrees, one could argue that there is a *minimal threshold*, below which there is no freedom of will at all, and a maximal threshold, above which it does not make sense to talk of more or less free will. This position clearly does have some justification. We would not want to say that, beyond a certain level, being more conscious or clever would give an individual even more freedom of will. On the other hand, it is possible that a super-being, who is less physically and mentally constrained than humans, may feel that humans' natural limitations do reduce their freedom of will. This super-being's maximal threshold would therefore be set higher than ours. However, even accepting the existence of minimal and maximal thresholds, our examples should have made clear that there is a wide gray area between those thresholds where free will comes in degrees. Often a limitation or constraint doesn't remove freedom but reduces it, and those limitations or constraints can be overcome with effort and planning. In a wide variety of situations, freedom of will should be, and generally is, calibrated by degrees.

Follow-Up Questions

Adina Roskies

This is more a comment than a question: it may be useful to make clear that a view that holds that free will is dependent upon having certain capacities (a capacitarian account) naturally leads to the idea of degrees of free will, since capacities are graded. That is still consistent with the notion that there may be thresholds, below which something does count as having free will at all.

Hans Liljenström

It seems quite apparent that, in general, free will comes gradually for the reasons mentioned, but would it not also depend on the level of consciousness? From an evolutionary or developmental point of view, would it not be reasonable to assume that animals and humans evolve and develop increasingly complex behaviors, and hence perceive larger sets of choices, as they become more and more conscious or aware of several optional actions? Would that not reflect an increasing degree of freedom? For example, a snake might just "choose" between fleeing or attacking when a sound is heard or the ground vibrates, while a dog may consider a much broader spectrum of choices, supposedly without either of them being conscious of their intentions. Isn't it reasonable to assume that the dog still has a greater freedom of will than a snake? Likewise with a human being (supposedly with a more developed consciousness than other animals), who as a child may just perceive two choices, do or don't, while an adult normally would be aware of a large set of alternative choices. Could you not say that the degree of freedom increases with the number of perceived alternative choices, which presumably increases with the level of consciousness?

Deniz Arıtürk

Are all external interferences with an agent's ability to act in line with their intentions equally hard to resist and equally restrictive? For instance, is a criminal who acts violently and has an identifiable brain anomaly any more or less morally responsible than one who does not have any identifiable brain anomaly but has experienced sociological hardships like abuse, trauma, poverty, or discrimination? And given that our brains are constantly altered by and alter our environments, is the internal/external distinction meaningful?

Replies to Follow-Up Questions

Jonathan Hall and Tillmann Vierkant

These questioners seem to agree that free will comes in degrees. This may be because they accepted our account of free will as dependent upon having certain capacities (a "capacitarian account," as Roskies calls it) which naturally leads to the idea of degrees of free will, since capacities are graded. Their questions were therefore focused on whether other elements of human cognitive abilities also lead to degrees of freedom. These elements included complexity and consciousness (Liljenström). A further question was how to compare degrees of freedom across different constraints, such as brain damage vs. sociological hardship and whether the internal/external distinction is meaningful at all (Arıtürk).

Let's begin with this last point. We would argue that the internal/external distinction is meaningful. While we fully agree that our cognitive processes are tightly interwoven with the environment, it still seems very plausible to say that the rational machinery itself is completely (or at least to a very large degree, if one is convinced by the extended mind) inside the skull. It therefore makes sense to distinguish between constraints that tamper with the machinery itself and constraints that arise from inputs to the machine. This raises the broader question of how to compare losses of freedom from the reduction in different capacities. We do not try to answer that question in this paper and remain open to the idea that they may be incommensurable.

We were aiming to be inclusive in our text so that most philosophical accounts could agree with our answer. But as we make clear, some accounts will put greater emphasis on certain individual elements than others. Consciousness (mentioned by Liljenström) is a prime example of this. There are philosophers who think that a very robust consciousness is necessary for full human freedom, while others think that, e.g., the normal functioning of the rational machinery is key. There are real differences here, but also some open questions. For example, it might well be that the ability to understand and decide on complex problems is a key feature of human freedom, while remaining unclear as to how important conscious awareness is for this ability.[4]

Finally, regarding complexity (also mentioned by Liljenström), our account focused mainly on degrees of human freedom, so normally functioning rational machinery should be interpreted as normally functioning human rational machinery. As alluded to in our discussion of a hypothetical super-being, other entities of greater (or lesser) complexity may be subject to less (or more) constraints. In our framework, it is possible that some sophisticated animals could exhibit enough functional flexibility to be considered partially free, while others would fall below the minimal threshold.

---

[4] See Chapter 13 by Bayne on consciousness and freedom and Chapter 25 by Gavenas et al. on conscious control of action.

PART I, SECTION III

QUESTIONS ABOUT SCIENTIFIC EVIDENCE

8

How can we determine whether or not we have free will?

Alfred R. Mele

If everyone agreed about what "free will" means, we would have a definite starting point for answering this question. But such agreement does not exist. So what are we going to mean by "free will" when we try to determine whether or not we have it? I will work my way up to a suggestion.

Suppose we say that free will is the ability to act freely. Then we can treat free action—including, importantly, free mental actions like freely *deciding* to write a brief essay on free will—as the more basic notion and try to define free will in terms of it.

So what is a free action? Here, I limit myself to the two kinds of account of free action that may be most attractive on theoretical grounds. And because deciding (or choosing) to do something is the kind of action that receives the most attention in the philosophical literature on free will, I concentrate on free decisions.

Consider the following proposal:

> (P1) If a sane person consciously makes a reasonable decision to do something on the basis of good information and no one is pressuring him, that person freely decides to do that thing.

If these conditions are sufficient for acting freely, and free will is the ability to act freely, then evidence that sane people sometimes consciously make reasonable decisions on the basis of good information without being pressured by anyone would be evidence of the existence of free will. (Note: P1 is a proposed *sufficient* condition for freely deciding to do something, *not* a collection of *necessary* conditions for this. P1 does not say, for example, that Ann cannot freely decide to

Alfred R. Mele, *How can we determine whether or not we have free will?* In: *Free Will*. Edited by: Uri Maoz and Walter Sinnott-Armstrong, Oxford University Press. © Oxford University Press 2022. DOI: 10.1093/oso/9780197572153.003.0008

write an essay on free will while experiencing some pressure from her chairperson to write it.)

Some people regard P1 as too weak to take us all the way to *freely* deciding. What is missing, according to them, is an ability to decide otherwise, on a very specific interpretation of that ability. Here is a way to picture what they have in mind. Suppose that at time *t* (in the actual universe) Ann decided to accept an invitation to write an article on free will. If every possible universe that is exactly like the actual universe in all respects up to time *t* includes Ann's deciding to accept that invitation at *t*, then Ann's universe is deterministic and, according to the critics of P1 at issue now, she was not able to do otherwise at *t* than decide to accept that invitation. But if, in some possible universes that are exactly like the actual universe up to *t*, Ann decides at *t* to reject the invitation, then she was able to decide otherwise. And if, in some such possible universe, Ann *reasonably* decides at *t* to reject the invitation, then she was able to decide otherwise *reasonably*. If we think of decisions as having causes (as I do), one thing that is needed for the ability to decide otherwise and for the ability to decide otherwise reasonably, according to critics of P1, is that the connection between Ann's decision to accept the invitation and its proximal causes is not deterministic—that it allows for the possibility of Ann's not making that decision at *t*. Another thing that is needed, according to these critics, is that the state Ann is in at the time does not preclude her making an alternative decision then (as opposed, for example, to simply remaining undecided) and does not preclude her making an alternative reasonable decision then. In that case, her state at the time includes a potential cause of her deciding to reject the invitation to write the article.

This brings us to the following revised proposal:

(P2) If a sane person consciously makes a reasonable decision to A on the basis of good information, no one is pressuring him, *and that person was able to decide otherwise in the sense just explained and, in particular, reasonably decide otherwise*, that person freely decides to A.

P2 is P1 plus the italicized bit. What would count as evidence that people make free decisions according to P2? I already commented on evidence for the unitalicized bit. So we are down to the ability to decide otherwise reasonably, in the sense explained. What is called for here is evidence that any laws linking decisions to their proximal causes are probabilistic and leave room for alternative reasonable decisions. If this sounds like a tall order, it should. Finding evidence that any laws linking decisions to their proximal causes are deterministic and do not leave any room for alternative decisions (reasonable or otherwise) is also a tall order. (Note: P2, like P1, is a proposed *sufficient* condition for freely deciding to do something.)

It is sometimes claimed that we never *consciously* decide to do anything—that all of our decisions are made unconsciously.[1] Evidence for that is evidence that we never satisfy the conditions that P1 and P2 say are sufficient for deciding freely. (Notice the word "consciously" in P1 and P2.) Also, any evidence for the thesis that the laws that govern decision-making are deterministic would be evidence that we never satisfy the conditions for freely deciding set forth in P2.

I should add that some philosophers (e.g., Harry Frankfurt and Robert Kane) believe that there is more to free will than the ability to act freely, even when it is made clear that deciding or choosing to do something is itself an action. There are too many ideas about what more there is to take up that issue here. But even if free will goes beyond the ability to act freely, researchers who set their sights on free action—including, importantly, deciding or choosing freely—will find much of interest to investigate.[2]

Follow-Up Questions

Mark Hallett

In discussing decision-making, aren't you just arguing that each potential cause is probabilistic? Then, if there are multiple possible courses of action, the outcome could be any of the possibilities in any

---

[1] See Chapter 12 by Block on consciousness.
[2] The author suggests *Free: Why Science Hasn't Disproved Free Will* by Mele (2014) and "Free Will" by O'Connor and Franklin (2020) for further reading.

circumstance. Most, but not all, of the time, if you knew the weights of the probabilities, you could predict the outcome.

Also, in your example, when you say Ann is making a decision, what do you consider Ann to be?

Silvia Seghezzi and Patrick Haggard

You propose a sufficient condition for freely deciding to do something. But to what extent does being "well informed" contribute to the sufficiency of this proposal? What is the added value of making a decision on the basis of good information rather than bad information on the voluntariness of the action? Can we not describe as "free" a decision that has been made based on bad information?

Hans Liljenström

With regard to the example of Ann agreeing to write an essay on free will, what if her supervisor, friend, parents, or whoever has an expectation that she should write it, and she feels obliged to do it, not because she wants to but because she perceives this expectation? Does she then act freely, if she agrees to write it, under (perhaps unconscious) pressure?

Replies to Follow-Up Questions

Alfred R. Mele

I wish I could answer all the interesting questions raised by my colleagues, but I lack the space to do that here. I hope readers find the following responses useful.

I offered two different (but overlapping) sets of sufficient conditions for freely deciding to do something. One set includes probabilistic causation, as Mark Hallett says, and, for the uninitiated, I described a way of conceptualizing probabilistic causation that is popular in the philosophical literature on free will. Yes, if you knew the weights of the probabilities for the decisions open to a person at a time, you could make predictions based on those weights, and an undetermined, predicted decision can be freely made according to proponents of P2.

Since my answer to the preceding question is brief, I'll answer Hallett's other question here too: "What do you consider Ann to be?" Imagine that Ann is a professional ballerina. She is performing on stage. You have a front-row seat, and you're watching her every move. I consider Ann to be the whole human being you're watching. Some people like to say "You are your brain." But I can't go along with that—and not for any fancy philosophical reasons. Ann's brain weighs about three pounds, and she weighs a lot more than that.

Silvia Seghezzi and Patrick Haggard ask whether a decision based on bad information can be free. Yes, in principle. But consider the following case. All of old Aunt Jo's money will go to her beloved nephew, John, when she dies. Jo is suffering from COVID-19, and John needs money in a hurry. John fabricates evidence that a certain cocktail that includes chloroquine and Clorox will cure Jo, knowing that, actually, it is likely to kill her. On the basis of John's skillful presentation of the fabricated evidence and her trust in him, Jo decides to inject herself with the cocktail John prepared for her; she acts on that decision and she dies as a consequence. As some people see things, John's deception of Jo renders Jo's decision unfree. If, in your opinion, Jo freely decided to inject herself with the mixture, you are free to regard the bit about being well informed in P1 and P2 as a useless part of the sufficient condition offered.

In response to Hans Liljenström's question about Ann, one thing one should do is distinguish wanting something for its own sake from wanting something as a means to an end. Ann might want to write the essay as a means of doing what she feels obliged to do. Did I want to undergo the dental implant surgery I had a few years ago? Certainly not for its own sake. But I did want to undergo it as a means to pain reduction and to dental health.

Suppose that, out of a sense of duty, as we say, Ann decides to write the essay despite her regarding writing it as an unpleasant chore. Suppose that, to some extent, she is pressured by others to write it. In that case, does she freely decide to write it? Neither P1 nor P2 answers this question. Nor are they meant to answer it, of course. They are not definitions of deciding freely. They are merely (proposed) sufficient conditions for deciding freely. One might ask where, on a continuum that has zero pressure from others at one end and maximum pressure from others at the other end, we find a dividing point between free and unfree decisions. P1 and P2 aren't in the business of answering that question. They characterize allegedly clear cases of free decision-making from two different theoretical perspectives.

9

What kind of neuroscientific evidence, if any, could determine whether anyone has free will?

Adina L. Roskies

I have argued that no neuroscientific evidence that one could feasibly obtain could settle, *by itself*, whether anyone has free will (Roskies, 2006). One important reason for this is that the question of what it is to have free will is a philosophical question,[1] and thus decisions about philosophical issues, or philosophical commitments, play an important role in the answer. To illustrate, many people think that whether or not we have free will depends upon whether our brains are deterministic systems. However, that already involves a commitment that many philosophers will deny—that free will depends on whether determinism is true. There are plausible theories of free will upon which free will is independent of considerations of determinism.[2] However, let's suppose we accept that determinism is incompatible with free will (a philosophical position known as "incompatibilism"). Couldn't neuroscience tell us whether our brains are deterministic?[3]

"Determinism" is a technical term. For a system to be deterministic means that if you knew the complete state of the system at any time, and all the laws that govern its behavior, then you could completely specify the state of the system at any other time. I have argued that even were we to accept incompatibilism, neuroscience still could not reveal

[1] For an alternative perspective, see Kreiman's replies in Chapter 17 on the neural basis of the will and Chapter 30 by Kreiman et al. on computational models and free will.

[2] See Chapter 4 by Sinnott-Armstrong on freedom, Chapter 5 by O'Connor on free will, Chapter 6 by O'Connor on determinism, and Chapter 8 by Mele on free will.

[3] See Chapter 29 by Liljenström on stochastic neural processes.

Adina L. Roskies, *What kind of neuroscientific evidence, if any, could determine whether anyone has free will?* In: *Free Will*. Edited by: Uri Maoz and Walter Sinnott-Armstrong, Oxford University Press. © Oxford University Press 2022. DOI: 10.1093/oso/9780197572153.003.0009

whether our brains are deterministic or incorporate random processes (Roskies, 2006). The kind of neuroscientific information that we think might bear upon whether or not the brain is deterministic is either information that suggests that a certain pattern of brain activity is predictable according to causal laws or that it is unpredictable given other information we have. We use predictability as a proxy for necessitation or deterministic behavior, but in fact predictability is not a fail-safe guide to it. Other sciences show this: there are predictable but arguably indeterministic systems (e.g., macroscopic but fundamentally quantum mechanical systems—probably most objects we are familiar with) and unpredictable deterministic systems (e.g., deterministic chaotic systems, best exemplified mathematically since we do not know if physical systems that attempt to realize them are deterministic). More importantly, neuroscience itself does not give us enough or the right kind of information to weigh in on determinism. First, not enough: the brain is an immensely interconnected and complex physical system. In order to determine whether some seemingly stochastic event really is stochastic, we would have to know the complete state of the system at a time. But no realistic neuroscientific experiments can ever reveal the complete state of such a large and complex biological system. In the absence of complete knowledge, some event could always be attributable to causal consequences of an event in another part of the system. Second, not the right kind: what looks to be stochastic activity at one level of grain or spatial scale can be due to deterministic causal processes at a finer grain. This seems to be true down to the level of individual molecules, and perhaps lower still. So our knowledge of the state of the brain has to be knowledge at the fundamental level, which is the level of concern to physicists, not experimental neuroscientists.[4]

But perhaps we are asking too much of our neuroscience. Perhaps, rather than asking what could determine whether we have free will, we should ask what could provide evidence for or against free will, or determinism, or contribute to our understanding, given other

---

[4] On a more prosaic level, even if the brain were deterministic, neuroscientists could not even show that all brain events are determined because they study brain activity in a laboratory setting, missing real-world decisions on important issues. And, as in all science, measurements and theory are never completely accurate, which would be required to show determinism to be true.

commitments? Here the answer very much depends upon what our other commitments are. If, for instance, you believe that having free will depends upon having a certain array of cognitive and agential capacities, such as, for example, the capacity to act for reasons,[5] to control certain impulses, and so on, then you might try to discover what sorts of neural architectures support those kinds of capacities.[6] Once you have developed such an understanding (which would, of course, depend upon a lot more than just neuroscience), you might be in a position, on the basis of neuroscientific information, to determine that some agent has such capacities, or that some other agent lacks one we deem critical. Indeed, once we better understand the nature of these capacities, it might be easier to argue on the basis of neural data that someone lacks a capacity, for it may require less information to demonstrate that a system is broken than to determine that it is working properly. Notice, however, that the relevance of any of this neuroscientific information is highly dependent upon deeply philosophical commitments about the nature of free will, such as which capacities are needed.

Or suppose, like Libet (1985) and many others, one is committed to a conscious decision being a proximal cause of freely willed action. It is conceivable that one day we will understand what the biological hallmarks are of a mental state's being conscious and what the causal processes are that mediate between decision and action.[7] Perhaps then neuroscience would allow us to track the relevant causal processes and would enable us to identify instances in which the hallmark of consciousness was absent. I would bet, however, that if we found it was always absent, we would be more prone to think that the signal we had provisionally identified as the marker of consciousness was not the essential signal after all. For, in order to identify such a marker of consciousness to begin with, we would need to correlate it with a behavioral report, and since it seems to us that we often consciously decide things, it would be hard to see how we could get evidence that in processes leading to action consciousness is always

---

[5] See Chapter 15 by Sinnott-Armstrong on reasons.
[6] See Hopkins & Maoz's replies in Chapter 23 on beliefs and desires.
[7] See Chapter 12 by Block on causation by consciousness and Chapter 25 by Gavenas et al. on conscious control of action.

absent.[8] Neuroscience may provide defeasible evidence that some, but not all, actions arise independently of consciousness.

In order to determine what kind of neuroscientific evidence could contribute to understanding whether or not we have free will, we must have a sense of what free will requires. It is likely that our philosophical and neuroscientific views will coevolve. Understanding the relevance of various brain measurements for volition and action will involve developing highly articulated neuroscientific theories about mind and behavior. For example, Libet, Gleason, Wright, and Pearl (1983) argue that the timing of the readiness potential (RP) demonstrates that we do not have free will, and many since have focused on this neuroscientific measure as being central to the question.[9] However, work over the past decade has shown Libet's original argument to be flawed in several ways (Roskies, 2011, Schurger et al., 2021) and the RP to likely be a methodological artifact (Schurger, Sitt, & Dehaene, 2012). The misunderstanding of the relevance of the RP for free will is a cautionary tale, resulting from overly simplistic and unrealistic theories of what certain brain signals indicate. The deep lesson we should take away is that our interpretation of brain data and its relevance to a profound philosophical question are mediated by a lot of theory, both philosophical and psychological, and that our neuroscience is unlikely to uncover deep truths about metaphysical questions that don't presuppose some controversial metaphysical views. That is not to say that neuroscience will be irrelevant to determining which are our best theories, but its interpretation will not be independent of them.[10]

Follow-Up Questions

Uri Maoz

Roskies claims that no neuroscientific evidence that one could feasibly obtain could settle, by itself, whether anyone has free will because

---

[8] See Chapter 24 by Mudrik & Schurger on consciousness.

[9] See Chapter 19 by Bold et al. on arbitrary and deliberate decisions and Chapter 28 by Lee et al. on timing of mental events.

[10] See Chapter 26 by Hopkins et al. on consensus on consciousness and volition.

the question of what it is to have free will is a philosophical question. She then notes, correctly in my opinion, that many people think that whether or not we have free will depends upon whether our brains are deterministic systems. And she devotes much of the chapter to a discussion of the extent to which neuroscience could shed light on that question.

I would like to suggest another approach to the question. It is not very different from Roskies' but may help clarify the potential role of neuroscience in answering the main question of this chapter. The question of whether humans have free will can be broken up into two questions:

1. What would it take for humans to have free will?
2. Do humans possess whatever it is that it takes for them to have free will?

The first question, as Roskies notes, is a philosophical one. However, once an answer is given to that question—be it a nondeterministic brain, the ability to respond to reasons, and so on—we must answer the second question. Science in general is a good way to go about answering that second question. And to the extent that what we must possess pertains to the brain, neuroscience can help us answer the second question and thus help determine whether humans have free will. Thus, what kind of neuroscientific evidence could determine whether humans have free will then depends on the answer to the first, philosophical question.

Hans Liljenström

An incompatibilist view claims that free will requires indeterminism (at some level), but many would argue that it is a necessary but not a sufficient condition. Pure randomness, e.g., given by quantum effects at an atomic level, is indeterministic but has difficulty providing a basis for free will. Are there any philosophical views on which free will could be compatible with randomness?

Tzachi Rotbain and Liad Mudrik

Clearly, as the author writes, "no realistic neuroscientific experiments can ever reveal the complete state of such a large and complex biological system" as the brain. Might there be a way to narrow down elements in the state of the brain that are irrelevant to decision-making, and focus only on the relevant ones? And then, if we find an "if and only if" relation between my decisions and the state of this narrowed down mechanism, would that strengthen the claim against free will, based on neuroscientific findings? If this narrowing-down does not work, is there any way to find an empirical compromise or approximation to test the ability to act otherwise?

Deniz Arıtürk

If our interpretation (and creation) of brain data is mediated by our philosophical and psychological theories, how can brain data ever provide an objective metric by which to adjudicate among competing theories?

Replies to Follow-Up Questions

Adina Roskies

I have argued that whether or not we have free will is not a purely empirical question. As Uri Maoz has suggested, the question can be fractionated into two:

1. What would it take to have free will?
2. Do humans possess that which it would take?

The former is philosophical, and the latter empirical. I concur, although I also believe that philosophical theorizing can and ought to be informed by science, so that even the answer to (1) is not completely independent, at least methodologically, of scientific inquiry. Moreover, our interpretation of scientific evidence itself is mediated by philosophical commitments, so scientific inquiry is not philosophically neutral.

As Maoz points out, many people believe that having an indeterministic brain is an apt answer to (1) and that neuroscience can simply settle the question of whether the brain is deterministic or indeterministic. I disagree, for several reasons. Because of the complexity of the brain and ineradicable noise, I have argued that no neuroscientific techniques can empirically settle the issue of determinism in the brain.[11] Moreover, because there are both predictable systems that are (arguably) indeterministic (e.g., macroscopic quantum systems) and unpredictable systems that are deterministic (e.g., deterministic chaotic systems), predictability alone is not a sure guide to determinism (Roskies, 2006). Finally, seemingly erratic behavior at one level of description can turn out to be law-governed behavior at a finer-grained level of description. Because neuroscience operates at many levels of description, it is only at the most basic level that our techniques will have the resolution they need to answer the question of determinism. Thus, if there is a scientific answer to that question, it is likely to come from physics, not neuroscience. However, because we have competing theoretical accounts of the fundamental physical level as either fundamentally indeterministic (traditional quantum interpretations) or deterministic (hidden variable theories), it seems likely that even at the level of fundamental physics, there is still no clear and established answer to this question.

My objections to the view that neuroscience will not settle the free will question go deeper. Suppose the brain were indeterministic. Would that indeterminism be able to support free will? On first glance, indeterminism seems an unlikely ground for freedom. As Liljenström notes, randomness does not seem to provide a good basis for free will. It would be like predicating freedom (and responsibility) on the roll of a die—something that is up to chance, not us. There are some libertarian theories of freedom which nonetheless hinge freedom on randomness; two views that come to mind are agent-causation views, in

---

[11] See Chapter 29 by Liljenström on stochastic neural processes.

which the agent somehow (!) breaks the causal chain and inserts indeterminacy into the world at points of decision, and Kane's Libertarian view, in which select decision-events wherein opposing reasons are equally matched in strength are resolved by quantum indeterminacies in ways that are still compatible with agents authoring their actions. Although perhaps intuitively compelling, both of these theories seem to fail under scrutiny, for decisions seem unmoored from their makers: the person is not the real source of the decision. Interestingly, the attractiveness of these views lies in the elements that *causally tie* the ensuing action to the agent, e.g., the arguments Kane makes for why the decision should be attributed to the agent, which are the deterministic parts of the theories.

All these considerations have led me to believe that the question of determinism vs. indeterminism is a red herring. If free will is supposed to ground responsibility, then it should be based upon considerations other than randomness. In answer to Liljenström, I believe that randomness can be a part of a free will picture, but not the part upon which free will hinges. Suppose, for example, that we have multiple redundant neuronal pathways that can implement the same behavior (e.g., any of a number of neurons or neural populations can suffice to produce a movement). Activation of one rather than another of those neural populations may be random for a given action, and the behavior may still be an exercise of free will. But that this behavior is executed is causally dependent upon upstream signals, and it is this dependence that makes the behavior an action of the will. In sum, I don't think the question of free will hinges on the truth or falsity of indeterminism.

Some have expressed worry that if the answer to the question of free will is philosophical, then neuroscience will have nothing to contribute. I think those worries are misplaced. As we further investigate the processes of volition and come to better understand the architecture and dynamics of our neural systems, we will also better understand the functional roles of various elements and, in conjunction with our philosophical views, will be able to narrow our views about which brain states are relevant. I expect that our philosophical views and our scientific pictures will co-evolve: the science may suggest a functional picture we have not yet considered, and the philosophical views may inform the interpretation of the data.[12] As I have argued (e.g., Roskies, 2014), even if we find an "if and only if" relationship between decisions and brain states (as mentioned by Rotbain and Mudrik), this will not settle the question against free will, but it may make a particular compatibilist or incompatibilist view seem to most accurately describe how we work: we ought to be sensitive to goodness-of-fit between theory and data. Again, I caution against putting too much weight on the question of determinism/indeterminism and instead think that what we ought to focus on are functional roles, the delineation of capacities, and their interactions. Both philosophy and neuroscience will be important for this process.

Some of the respondents (especially Arıtürk) seem to worry that if the question of free will is even partly philosophical, then the answers lack objectivity or scientific value. Here is not the place for a lecture in philosophy of science, but it seems worth reminding or informing the reader that all scientific work is philosophically laden: the idea that *objectivity requires that there is a scientific truth that makes no philosophical commitments* is terribly naïve. What is interesting about the free will problem is that it is an example of a scientific question that forces us to become aware of and confront philosophical commitments that might otherwise remain implicit or hidden, and it brings to the fore the inextricable entanglement of philosophy and empirical science.

[12] See Chapter 26 by Hopkins et al. on consensus on consciousness and volition.

10

What kind of behavioral experiments, if any, could determine whether anyone has free will?

Tim Bayne

There are three very different ways of approaching this question. The first way can be explored by asking how we might tell whether an AI system has free will.[1] Suppose that we encounter the sentient computer HAL 9000 from the movie, *2001: A Space Odyssey*, and we want to know whether HAL has free will. (Perhaps we are trying to work out whether to hold HAL responsible for what it does, and we think that the question of free will is relevant to that issue.)[2] Would behavioral data help us to decide whether HAL has free will? And if free will comes in degrees,[3] would behavioral data help us to decide *how much* free will HAL has?

It seems plausible to think that free will might come in degrees. After all, we appeal to cognitive capacities in distinguishing agents who have free will (such as neurotypical adult humans in states of unimpaired consciousness) from agents who don't (such as young children, non-human animals, and adults who are suffering from serious brain damage or are in altered states of consciousness). In asking whether an agent has the capacity for free will, we consider such things as whether they understand the implications of their actions; whether they are able to revise their intentions in light of new information; whether they are able to control their bodily movements; and whether they are able

---

[1] See Chapter 11 by Hall & Vierkant on artificial intelligence.
[2] See Chapter 14 by Yaffe on responsibility.
[3] See Chapter 7 by Hall & Vierkant on free will in degrees.

Tim Bayne, *What kind of behavioral experiments, if any, could determine whether anyone has free will?* In: *Free Will.* Edited by: Uri Maoz and Walter Sinnott-Armstrong, Oxford University Press. © Oxford University Press 2022. DOI: 10.1093/oso/9780197572153.003.0010

to inhibit their impulses. These capacities are relevant to the question of whether HAL has free will, and the question of whether HAL has these capacities can be addressed via behavioral methods.

Let's turn now to a second interpretation of the question—which is what I suspect the questioner really has in mind. On this interpretation, what's being asked isn't whether there are behavioral methods that could give us the same kinds of reasons that we already have for thinking that you and I have free will, but whether there are behavioral methods for deciding whether you or I *really* have free will. This type of questioner is fully aware that we ordinarily take ourselves to have free will, but they want to know whether that stance is ultimately justified. Perhaps—the skeptic might suggest—free will is an illusion even when it comes to neurotypical adult human beings who seem to be in full possession of their wits.

Whether behavioral data could address this kind of skeptical worry depends on what exactly grounds it. We would need to know why our skeptic harbors the suspicion that ordinary folks might not really have free will. Does the skeptic think that free will requires a "substantial self"—something that acts as an "oomphy cause" and transmits energy or momentum from an immaterial mind to the body?[4] Do they think that free will requires libertarian freedom—a kind of capacity to do otherwise that would be at odds with determinism?[5] Or do they think that free will requires consciousness to be some kind of non-physical property? If free-will skepticism is grounded in worries of this kind, then it is difficult to see how behavioral data might address it. After all, these concerns are fundamentally metaphysical, and they are unlikely to be amenable to direct behavioral evaluation. Consider, for instance, the suggestion that free will requires libertarian free will. What behavioral test could be conducted that would bear on the question of whether human beings have libertarian free will? It is hard to think of one. At the very least, the kinds of worries that animate most free-will skeptics are not easily resolved by behavioral data.

---

[4] See Chapter 2 by Hieronymi on the will.
[5] See Chapter 4 by Sinnott-Armstrong on freedom, Chapter 6 by O'Connor on determinism, and Chapter 8 by Mele on free will.

Finally, let's consider a third approach to this question. Begin with the thought that free will is a "folk concept," a concept that is employed in everyday contexts. As such, the question of whether Jane has free will (or carried out a certain action freely) can be understood as the question of whether a certain folk concept applies to Jane. Now, folk concepts are precisely the kinds of things that ordinary people ("the folk") have mastery over, and we can begin to chart the contours of the folk notion of free will by employing questionnaires and other behavioral methods. This is precisely what theorists working in the field of experimental philosophy have begun to do over the past two decades (e.g., Nahmias, Morris, Nadelhoffer, & Turner, 2005; Roskies & Nichols, 2008). Of course, knowing what ordinary people mean by "free will" won't itself settle the question of whether Jane has free will, but it surely represents an important and necessary first step in that process.

Follow-Up Questions

Liad Mudrik

If #1 is not what the question is all about, and #3 cannot settle the question of whether Jane has free will, and if #2 cannot benefit from behavioral experimentation, I guess the answer to the question discussed in this chapter is no: behavioral experiments cannot settle the question of whether someone has free will (or contribute to the free-will debate). Right? If so, could neuroscientific studies help here? If yes, why would they be superior to behavioral studies (given that our neural measures are often proxies for or correlations of behavioral phenomena)? If not, should we conclude that no empirical data could determine whether someone has free will?

Tomáš Dominik

Several parts of this answer touch on an underlying question, which I consider to be fundamental to the discussion we are having in this book. What evidence do we have to believe there is something to be called "free will"? The most obvious answer to my question might be that we believe we have free will because it is our everyday introspective impression that we are the authors of our actions. But is there any other evidence for the existence of free will besides introspection? And is our introspection adequate as evidence on this issue?

Replies to Follow-Up Questions

Tim Bayne

Liad Mudrik challenges me to come clean on the relationship between behavioral studies and free will. Consider Jane, our representative individual: could behavioral data tell us whether or not she has free will?

The first question I would want to ask is what we already know about Jane. Let's suppose that her psychological profile is completely unknown to us. In that case, there are many behavioral experiments that we could run which would have a bearing on whether we think that Jane has free will (better: on how much free will we think she has and/or the conditions in which we think she is in a position to exercise her capacity for free will). We could check whether she can understand the consequences of her actions, inhibit her urges, respond to reasons in appropriate ways, and so on. The answers to these questions (and many others that could be probed by behavioral experiments) are surely relevant to whether Jane has free will.

But suppose we already know that Jane is neurotypical. We know that she can understand the consequences of her actions, inhibit her urges, respond to reasons, and so on. In that case, what *further* information might we glean from behavioral studies—or indeed empirical studies of any kind—that could bear on the question of free will? I can imagine a lot of interesting answers to this question, but I don't know of any that don't rest on highly controversial assumptions about the nature of free will.

Here's a parallel question: Could behavioral data bear on the question of whether Jane is *rational*?[6] Certainly, for we might discover that

---

[6] See Chapter 1 by Yaffe on intention and Chapter 15 by Sinnott-Armstrong on reasons.

Jane has systematic deficits in (say) practical reasoning and that she can't work out whether to put water in the kettle first and then boil it or vice versa. But suppose we know that Janes has no such deficits and that she is a neurotypical human being. She's not particularly good at doing logic in her head (who is?), and she's prone to committing the conjunction fallacy (who isn't?), but by and large she can reason her way through most problems like the rest of us. In such a situation, I suggest, it's not clear what behavioral studies one might run to show that Jane is (or isn't) rational. Of course, there is a great deal one can learn about how Jane is reasoning that might measure up against the norms of (say) decision theory, but any attempt to use that data to pronounce on her rationality (or lack thereof) presupposes controversial assumptions about the relationship between rationality and the norms of decision theory.

In sum, although behavioral data can be brought to bear on the free-will debate, it seems to me that attempts in this direction will typically—and perhaps always—rest on contentious claims about the nature of free will and what it requires. Behavioral studies aren't a waste of time, but any attempt to draw a straight line between them and claims about free will is unlikely to succeed.

Tomáš Dominik asks a very good question about our grounds for believing in the existence of free will. I find it useful to break the issue down into two parts—a genetic question and an epistemic question. The genetic question is this: *Why* do people believe that they have free will? The epistemic question is this: Is the belief in free will well-founded? These two questions are intimately related, for the epistemic status of belief in free will depends on its grounds (Bayne, 2017).

In my view there are three answers to the genetic question that are worth considering. The first appeals to introspection: we believe that we have free will because free will (or at least something like it) is encoded in the contents of introspection. The second appeals to the structure of folk psychology. The beliefs and desires that we ascribe to each other under-determine what an agent will do on any given occasion, and we thus invoke some additional explanatory factor—a "freely willed decision"—to take up the slack between the psychological inputs to behavior and behavior itself. A third explanation focuses on moral considerations: moral accountability is unjust unless people act freely, and since we're committed to the practice of holding people morally accountable, we have strong reason to embrace belief in free will (Nichols, 2014). I don't know which of these hypotheses is correct; perhaps there is some truth in all three.

I don't have the space to consider the epistemic implications of each of the three hypotheses just considered, but let's consider the appeal to introspection. Would belief in free will be well-founded if it were based on introspection? One might say yes on the grounds that introspection is generally (although not, it must be admitted, universally) regarded as a reliable source of information. If I trust introspection when it tells me that I have a headache, should I not also trust it when it tells me that I am acting freely? On the other hand, one might plausibly argue that the question of free will is not the kind of question on which introspection could possibly deliver a trustworthy verdict. (Suppose you think free will requires being a self-caused cause—how could introspection generate reliable information about *that*?) The issues here are complex, for whether free will is the kind of thing that can be encoded in introspective content turns on the question of what free will itself involves. I leave as an exercise for the reader the task of identifying the epistemic implications of the other two answers to the genetic question.

11

Can a robot with artificial intelligence have free will?

Jonathan Hall and Tillmann Vierkant

To answer the question of whether an entity such as a robot with artificial intelligence (AI) can have free will, we will lean on our chapter where we asserted that to have free will is to have *the capacity to form and act in line with intentions*.[1] An entity with this capacity we will call an intentional agent, allowing us to reframe the question of the title as "Can a robot with AI or even the disembodied AI in a computer be an intentional agent?"

To be an intentional agent is to interact with the environment in a goal-directed way, consistent with the agent's representational and motivational attitudes to certain propositional content. By logically combining these intentional states, the agent forms goals that drive behavior. For example, an agent may rationally form the intention to open the fridge if she believes it contains beer and she (really) desires a beer.[2]

If the capacity to rationally combine intentional states to form intentions is necessary and sufficient for free will, then determining whether robots have intentional states would help answer our question. Presumably, we can already imagine a scenario where in certain narrowly defined domains the actions of a robot require intentional explanations, but in all known cases so far robotic abilities across a wide range of domains remain far below human standards. Clarifying that our aforementioned *capacity to form and act in line with intentions*

---

[1] See Chapter 7 by Hall & Vierkant on degrees of free will.
[2] See Chapter 1 by Yaffe on intention and Chapter 15 by Sinnott-Armstrong on reasons.

Jonathan Hall and Tillmann Vierkant, *Can a robot with artificial intelligence have free will?* In: *Free Will.* Edited by: Uri Maoz and Walter Sinnott-Armstrong, Oxford University Press. © Oxford University Press 2022. DOI: 10.1093/oso/9780197572153.003.0011

must be domain-general on a human scale will get us the result that robots even with AI are unlikely to be considered free agents in the next couple of decades, but beyond that it would be brave to rule out that robots could ever meet that criterion.

Assuming that one day the attribution of domain-general intentionality is necessary to explain their behavior, would that be sufficient to conclude that robots have free will? Intuitions might begin to diverge here, but for many the answer might still be no. One justification for the persistence of the skeptical intuition is that attributing intentionality on the basis of externally observable actions leaves open the possibility of confusing a simulation of agency with the real thing.[3] This mirrors a debate sparked by Turing (1950) in his paper "Computing Machinery and Intelligence."

Turing proposed that instead of asking the question "Can machines think?," one should consider whether there are imaginable digital computers which would do well in what he described as the "imitation game." In this game an interrogator asks questions to both a machine and a human and, from the answers, tries to determine which is which. John Searle (1980) famously argued through his Chinese Room thought experiment that the kind of proficiency, in converting inputs to outputs, shown by an entity passing the Turing test does not imply intentionality. In particular, Searle argued that if he was in a room receiving questions in Chinese and diligently followed English instructions regarding how to respond with Chinese characters, then he could appear to an interrogator to be conducting an intelligent conversation even though "I don't speak a word of Chinese."

Although there has been much debate about these thought experiments, those in the Searle camp insist that one cannot infer intentionality from behavioral output. A response is needed if one is ever going to be justified in attributing agency to robots. There are a number of forms that this response could take.

First, Searle is very clear that some machines can think, because, after all, "we are precisely such machines." Intentionality, in his opinion, must therefore be substrate-dependent, and it happens to be

---

[3] See Chapter 10 by Bayne on behavioral experiments.

the case that humans are made of the right stuff. It then becomes an empirical question as to whether other substrates can support intentionality. Theoretically, an alien could have or a robot could be built with a chemical and physical structure that supported intentionality (although it is not clear how we would know). Here we must differentiate between a moving and sensing robot with AI and the disembodied AI in a computer. In Searle's opinion, substrate-independent computational operations on formally specified elements are never sufficient for intentionality.

Second, a powerful response to Searle is that he is making an unjustified step from the fact that certain parts of a system don't understand Chinese to the assertion that the system as a whole does not. Even if Searle doesn't understand a word of Chinese, it is not obvious that the combined "Searle+instructions" system does not. No one would argue that individual neurons in the brain understand this essay, but hopefully the neural system as a whole does. Although, in our opinion, this objection is powerful enough to allow that robots with AI could be intentional agents, it is unlikely to persuade the skeptic.

Also plausible is a third approach, which Clark (1991) calls "microfunctionalism." This approach agrees that formal operations at the coarse-grained symbolic level of the Chinese Room experiment can't produce intentionality, but argues that formal relations at a fine-grained sub-symbolic level could provide the right kind of structure to support flexible domain-general behavior and the attendant emergent properties associated with intentionality. In this model, the cognitive system is assumed to be "at root a sub-symbolic system" that is scaffolded by intentional states at a higher level. If this is correct, then it is not the substrate that matters but the sub-symbolic system. This provides a framework in which it could be legitimate to claim that a robot with AI is an intentional agent.

There is not enough space in this essay to adjudicate among these three positions, but our personal hunch is that it seems much more likely that intentionality and free will are to be found in the organization of the substrate rather than the substrate itself, which makes options 2 and 3 the most likely contenders. Both of these are consistent with the possibility that robots with AI have free will.

Follow-Up Questions

Mengmi Zhang

What are the relations between intentionality and free will? Do they refer to the same thing? In theory of mind in philosophy, mental states like beliefs, desires, and perceptual experiences have intentionality in the sense that they represent or are about things or states of affairs. Does free will refer to a specific case of intentionality?

Deniz Arıtürk

Would robots with AI ever pass your requirement of "no external interference?"[4] If so, how? What, if anything, distinguishes the external interference on the actions of robots with AI (namely, that they are built by humans) from external interferences on the actions of humans (such as that their genes are a product of natural selection)?

Antonio Ivano Triggiani and Mark Hallett

Modern technologies show that the creation of an artificial brain is possible. Still, it's hard to reach a high degree of connections similar to that among human neurons, and it's also difficult to emulate their plasticity. So, does free will in a machine depend on its complexity?

Gabriel Kreiman

I really enjoyed reading this lucid answer. My question is, basically, what sort of *empirical data* would lead us to think that a robot has or does not have free will?

---

[4] See Chapter 7 by Hall & Vierkant on free will in degrees.

Let us start with behavior and consider a Turing test for free will. In room A, there is either a machine or a human; in room B there is either a machine or a human. You can ask *any* question. Based on the answers, how would you determine which room has an agent with free will?

I suspect that you will not like this formulation of the question. You may argue that behavior is not enough; there has to be "intentionality." I would like to make sure that the definitions connect to *empirically observable variables*. If behavior is out (or insufficient), then I will allow you now to record the activity of every single neuron in a biological agent and the voltage of every transistor in the robot. Feel free to add whatever variables you think are relevant here—calcium concentration in the pre-synaptic terminal, the position and composition of every atom. If there is an empirically measurable variable that you want, you got it! Based on those measurements, how would you determine which room has an agent with free will?

If the answer to these questions is that there is no empirically observable variable that would ever determine which agent has free will, then "intentionality," "free will," and similar terms have little scientific value as we cannot falsify them, we cannot measure them, we cannot use one in any empirical way to assess the other, etc. Then it seems that we have made up a specialized vocabulary with no connection to the physical world. For example, I could argue that certain flies have wtx3xtw. If I build a robot fly, would it have wtx3xtw? Never! Because it would lack zy6yz! Can we measure zy6yz! No. Can we measure wtx3xtw? No. But I shall assert that flies with zy6yz certainly have wtx3xtw.

David Silverstein and Hans Liljenström

Can a deterministic system, like (presumably) a computer or a robot, have intentions or free will? Suppose actions from free will are based on freely determined intentions. If algorithms for intentions are modeled deterministically, how can goal-directed decisions from environmental inputs be based on free will? Will these not be predetermined? Some aspects of human experience may appear random and may partially drive the development of intentions. If modeled intentions have a random component, can that help represent free will, or does this just dilute it? To what extent are intentions driven by self-models?

Replies to Follow-Up Questions

Jonathan Hall and Tillmann Vierkant

A number of questions were concerned with the different meanings of "intentional" (e.g., Zhang; Silverstein & Liljenström). We are happy to clarify these. Intentions as executive states that drive behavior are a subgroup of the wider category of intentional states that also encompasses, for example, beliefs (often referred to as Brentano-intentionality or aboutness). As we say in the text, in order to form intentions in the former sense, it is necessary that an agent can rationally combine intentional states in the latter sense (including beliefs). The biggest obstacle for allowing that robots with AI might have free will is that they do not seem to have any intentional states in the latter, Brentano sense.

An additional problem is that even if robots behaved as if they had Brentano-style intentionality, there is a big debate on whether that means they really do have it. In a way, the same problem exists obviously for humans, but in our own case we have intuitive first-personal access, and we assume that other humans with very similar brains to our own will probably also have it. We do not want to judge how important this evidence is, but it explains the intuitive difference between robots and humans.

Given this clarification, it is now much easier to answer the other questions. In a way, we here have reduced the question of whether robots can have free will to the question of whether robots could have intentional states in the Brentano sense that they could rationally manipulate to form intentions. There is a large literature in the human case on whether the having of intentional states is compatible with having a designer (Arıtürk), or what kind or degree of complexity is required (Triggiani & Hallett), or whether there could be an empirically observable variable to test for the existence of this (Kreiman). All we can do here is say that the answers to these questions in the human case will allow us to answer the robot case as well.

The one specific claim that we can already deduce from this is that randomness is not a major factor for free will on this account (Silverstein & Liljenström) because it seems unlikely that randomness will play a major role in explaining intentionality in humans. This outcome is hardly surprising, though, because we started off with a compatibilist notion of free will in the first place. So our answer does not address whether robots could have incompatibilist free will, but it does ask interesting questions of the incompatibilist. Would a robot that possesses aboutness and the ability to form intentions in a rational way really be not free if randomness played no role in its decision-making? And would it change anything if we added a randomness generator to the robot that could influence its decision-making in some way?

PART I, SECTION IV

QUESTIONS ABOUT CONSCIOUSNESS

12

Do conscious decisions cause physical actions?

Ned Block

1. What is a conscious decision?

A decision in one sense of the term is the formation of an intention. A phenomenally conscious decision in one sense of the term is then a phenomenally conscious formation of an intention. To say that forming the intention is phenomenally conscious in this sense is to say that there is something it is like to form the intention, that the formation of the intention has a phenomenal "feel." That phenomenal feel can take the form of an awareness of making a choice. Or it can be a matter of being phenomenally conscious of the decision *as a decision*, in which case the subject must possess the concept of a decision. It seems, though, that there are plenty of conscious decisions that don't feel like anything at all. We make many minor choices every day. Does it always feel like something to make them? Perhaps there is only a feeling of choice when there is some kind of deliberation about the choice. Still, our mundane everyday choices are conscious in the sense of "access-consciousness."

Access-consciousness is immediate global availability to cognitive processing, whether that cognitive processing is or is not itself phenomenally conscious (Block, 2002). An access-conscious content is immediately available to reasoning, planning, evaluating, problem-solving, reporting, memory, and other cognitive processes. If there are Freudian repressed states, they are unconscious in the access sense whether or not they have any phenomenal feel.

I just introduced three senses of "conscious decision": (1) a phenomenally conscious decision, (2) a decision that is phenomenally

Ned Block, *Do conscious decisions cause physical actions?* In: *Free Will.* Edited by: Uri Maoz and Walter Sinnott-Armstrong, Oxford University Press. © Oxford University Press 2022. DOI: 10.1093/oso/9780197572153.003.0012

conscious as a decision, and (3) an access-conscious decision. Of these three senses, the first two involve a phenomenal feel. And it is the presence of the phenomenal feel involved in such conscious decisions that give rise to the problem to be discussed here. That problem is this: there is experimental evidence that has been taken to support the claim that phenomenally conscious decisions are "epiphenomenal" in the sense that they have no causal effects on bodily movements (Libet, Gleason, Wright, & Pearl, 1983; Passingham & Lau, 2006).[1] In these experiments, unconscious neural events leading up to both the decision and consciousness of it are alleged to be found *prior* to consciousness of the decision; indeed, some experiments suggest that the consciousness of the decision can occur at least in part *after* the action (Lau, Rogers, & Passingham, 2007). (Whether or not these experiments really do establish the conclusions they are taken to establish is not my concern here; my concern is what follows from these claims if they are true.) Some commentators conclude from this sort of evidence that the conscious decision to act is not causally efficacious in producing the action because the unconscious neural events are sufficient to cause the action. This reasoning is my target.

I will explain why this reasoning is mistaken in terms that apply to *all mental events*. I will begin with the example of conscious vision, because we understand the psychology and neuroscience of vision much better than we understand any other aspect of the mind. In particular, there are dramatic cases in which unconscious and conscious states have *conflicting* contents. The lessons from such examples will then be applied to the case of conscious decisions.

2. Are conscious perceptions epiphenomenal?

All conscious mental events, including conscious perceptions, involve unconscious processing. Visual perception, conscious and unconscious, is typically processed by the lateral geniculate nucleus and the first cortical visual area, V1, on the way to conscious processing in

---

[1] See Chapter 25 by Gavenas et al. on conscious control of action.

higher visual cortex.[2] (For some kinds of unconscious perception, the pathways involve the superior colliculus[3] and the pulvinar,[4] bypassing the lateral geniculate nucleus–to-V1 route.) There is good reason to believe that representations at the level of the lateral geniculate nucleus and V1 are not part of the neural basis of consciousness.[5]

The major theories of consciousness that are relevant to this issue agree that processing in the lateral geniculate nucleus and V1 *precedes* conscious processing. I will explain with respect to three of the major theories of consciousness.

(1) The global workspace theory dictates that conscious processing of a stimulus begins—at the earliest—270 ms after the stimulus, long after the stimulus is extensively processed in the lateral geniculate nucleus and V1 (Dehaene, Changeux, Naccache, Sackur, & Sergent, 2006). (2) Higher-order thought takes somewhat *more* time than global broadcasting, so theories of consciousness based on higher-order thought (Brown, Lau, & LeDoux, 2019; Rosenthal, 1986) also allow for substantial unconscious processing prior to conscious processing. (3) The recurrent processing approach to consciousness also dictates that conscious processing occurs well after stimuli are extensively processed in the lateral geniculate nucleus and V1 (Lamme, 2003; Pitts, Metzler, & Hillyard, 2014). According to Victor Lamme's version of the recurrent processing account, conscious perception requires processing in the lateral geniculate nucleus and V1, then processing in higher visual areas and then, finally, feedback to V1 (or V2). And there is independent evidence for the need for the feedback to V1 (Block, 2007; Silvanto, Cowey, Lavie, & Walsh, 2005). So on this account, conscious perception requires, first, unconscious activations in V1 and, then, a second round of activations in V1.

I conclude that with perception, as with decision, there is reason to accept that some *unconscious visual processing precedes all conscious*

---

[2] Brain regions (such as visual cortices V3, V4, and V5 and inferior temporal cortex) that process more abstract, high-level visual information. See Figure 2a in the Brain Maps section.

[3] Structure in the midbrain. See Figures 1b, 1c, and 3 in the Brain Maps section.

[4] A group of nuclei within the thalamus. See Figures 1b, 2a, and 3 in the Brain Maps section.

[5] Some of this evidence is summarized in Koch (2004) and Koch, Massimini, Boly, and Tononi (2016).

visual processing in the same stream. Nonetheless, we can agree that conscious perceptions are involved in the production of actions. Consciously seeing the red light is *often* part of the cause of stepping on the brake.

There is a complication, however. When one event causes another, some of the properties of the cause may be causally efficacious in producing the effect and others not. When the brick flying through the air breaks the window, it is in virtue of the brick's mass and velocity that the window breaks, not in virtue of its color: the color is causally inefficacious in breaking the window. The soprano's high C "Help!" may shatter the glass, but the meaning of the word "help" is not causally efficacious in shattering the glass (Dretske, 1988).

In terms of this distinction, then, how do we know whether it is in virtue of the *conscious aspect* of seeing the red light that I stepped on the brake? That is, how do we know whether the conscious aspect of seeing the red light is causally efficacious in producing the action of stepping on the brake?

A conscious perception has conscious and unconscious aspects, and when a conscious perception causes something, it will not always make sense to ask which aspects are causally efficacious. An iceberg displaces an amount of water equal to the weight of the whole iceberg, so it is the whole iceberg that is causally efficacious in that respect, not just the part below water. If the above-water part of an iceberg hits a ship, we cannot conclude that the below-water part was not causally efficacious, since without the below-water part there would be no above-water part to hit the ship. The same point applies to conscious mental events: without their unconscious part there would be no conscious part.

Still, in many cases we can ask whether the conscious part is causally efficacious, that is, whether it was at least partly in virtue of the conscious part that the effect happened. In some cases the answer is demonstrably yes.

In the light of controversies over whether to use the "objective threshold" (better than chance responding) or the subjective threshold (the subjects' belief that they see something), I will use the following methodology: if a perception is above both the objective and the subjective threshold, then the perception is conscious. The principle that if a perception is below both the objective and subjective threshold it is not conscious is also very plausible but not very usable since in many cases, subjects are sure they did not see the stimulus but can still answer questions at better than chance accuracy. A more useful principle for classifying unconscious perception is: if the subject is sure that they are unaware of the stimulus (or its properties) and if the putative unconscious content is incompatible with a conscious content, then the content is unconscious. Examples follow.

There is evidence that unconscious perception can influence behavior. In one experiment (Debner & Jacoby, 1994), subjects were presented with a masked word and then asked to complete a word stem, but not with the word they saw if they saw a word. (Masking—the presentation of a pattern before or after the original stimulus—can make the stimulus hard to consciously see if the timing is right.) If the word "reason" is presented consciously (lightly masked), then the subject can succeed in avoiding the presented word in completing the stem, for example, by completing "rea___" with "reader." But if "reason" is presented unconsciously (heavily masked), then the subject is more likely than baseline to complete the stem "rea___" with "reason." (There is an issue of whether the perception of "reason" that I described as unconscious is really unconscious as opposed to weakly conscious, but I cannot take up that issue here.)

Similar "opposite" effects of conscious and unconscious processing occur in other kinds of visual perception. For example, V1 can register black stripes on a white background that are too narrowly spaced to see consciously. If the density of such stripes is greater than 50 cycles per degree of visual angle, the stimulus looks to be a uniform gray field—as far as conscious vision is concerned. But stimuli that are substantially higher than 50 cycles per degree are registered by V1 as stripes (He & MacLeod, 2001). Similarly, if two colors (e.g., red and green) alternate at frequencies above 10 Hz, viewers consciously see a single fused color (in this case, yellow) rather than flickering different colors. But V1 and the lateral geniculate nucleus respond to flickering colors at frequencies up to 30 Hz (Gur & Snodderly, 1997).

The unconscious perception of stripes at higher than 50 cycles per degree—say, 60 cycles per degree—and unconscious perception of red and green flickering fit the principle mentioned earlier: if the subject is sure that they are unaware of the stimulus (or its properties) and if the putative unconscious content is incompatible with conscious content, then the content is unconscious. The content of stripes at 60 cycles per degree is one that the subject denies and is incompatible with the subject's certainty that they are seeing a uniform gray field. Similarly, the content red and green flickering is incompatible with the subject's certainty that they are seeing a uniform unflickering yellow.

Suppose a subject has the task of pressing the button marked "yellow" if the stimulus is yellow and a button marked "red & green" if the stimulus is flickering red and green. If the stimulus is a red/green flickering stimulus at 12 Hz, the subject will consciously see yellow (since the flicker rate is above 10 Hz) and so can be expected to press the "yellow" button.

But if the stimulus had been sufficiently degraded or masked so as to be entirely unconscious, the stimulus would have registered in unconscious processing as red and green flickering (since unconscious processing in the lateral geniculate nucleus and V1 registers flicker up to 30 Hz), so the yellow color would not have been perceptually registered. And if the resulting unconscious perception had an effect on behavior, it would incline the subject toward the "red & green" rather than the "yellow" button (Gur & Snodderly, 1997).

When the colored stimulus flickers at 12 Hz, resulting in the subject's pressing the "yellow" button, we can conclude that the conscious aspect of the processing was causally efficacious, since the earlier, unconscious part by itself would not have influenced the subject's behavior in the direction of the "yellow" button. Similarly, if the stripe density of a stimulus is 60 cycles per degree, a subject will classify it as uniform gray on the basis of conscious perception. But if the perception had had no conscious part, the visual system would have registered it as striped rather than uniform, so it would have inclined the subject toward the striped response—to the extent that the unconscious perception would have causally influenced a response. So, in the conscious case, we can conclude that the conscious aspect was causally efficacious.

The counterfactual test I am using has to be applied carefully. If the exposed part of an iceberg caused damage to a ship sufficient to sink it, we can ask what would have happened had the top part of the iceberg not been there so that the iceberg was entirely below water. The ship might have been sunk anyway, though through a different causal path. Still, if the result goes the other way—if the iceberg would not have sunk the ship had it not had the above-water part—then we can reasonably conclude in normal circumstances that the above-water part was causally efficacious.

The resulting picture of the relation between conscious and unconscious mental events is that when a conscious mental event is causally efficacious, we can sometimes ask whether it is causally efficacious in virtue of its conscious aspect. I have just given examples that show *the conscious and unconscious aspects can have different and opposed effects on behavior* in at least some cases. In these cases, it is particularly obvious that the conscious aspect of the mental event is causally efficacious.

But even in the case where the influence of the conscious and unconscious aspects of the mental events point in the same direction, they may make somewhat independent contributions to the behavioral effect. There is reason to believe that in vision, conscious and unconscious representations may have different contents. Representations in the lateral geniculate nucleus and V1 are always unconscious. They have "low-level" contents, zero-crossings, edges, orientations, contrast, textures. "High-level" contents, such as representations of faces, emotional expressions, causation, and numerosity, can be either conscious or unconscious, but to make some of these contents unconscious requires very brief presentations, highly degraded stimuli, or one or another kind of masking. Thus, while representations of low-level properties will typically have a substantial unconscious component, representations of high-level properties will often be conscious.

3. Back to phenomenally conscious decisions

Conscious decisions (I'm talking about phenomenally conscious decisions here) are conscious mental events, and so the points just made about *all conscious mental events* apply to them. If the subject is choosing between salad and chocolate cake, the unconscious aspect of the decision might incline the subject toward the chocolate cake, whereas the conscious aspect might incline the subject toward the salad. If the subject chooses the salad, then the conscious aspect was causally efficacious. However, if the conscious and unconscious aspects inclined the subject toward the same decision, *both* may be causally efficacious. With decision, as with perception, the unconscious and conscious aspects of the decision can point in the same direction but make somewhat independent contributions, in which case again the conscious aspects are causally efficacious. An unconscious part of a mental event always precedes conscious aspects, but the conscious aspects may nonetheless be causally efficacious.

With decision, as with perception, we can expect that there will be differences between the kinds of contents that will typically be unconscious and those that will typically be conscious. An unfortunate legacy of the Libet-style experiments[6] is a focus in the neuroscience of decision on very simple contents that can be either conscious or unconscious, basically go/no go contents. The field would be better off with an increased emphasis on the contents of decision and on which contents can be expected to be conscious and which unconscious.[7]

Follow-Up Questions

Mark Hallett and Liad Mudrik

As pointed out, unconscious events can cause physical actions, but in some circumstances it appears that the action seems to have depended on a conscious event. However, a conscious event is likely complex. Is the consciousness aspect of the event the critical part? Could the consciousness part be the red of the brick and an unconscious part be the mass of the brick?

---

[6] See Chapter 19 by Bold et al. on arbitrary and deliberate decisions and Chapter 28 by Lee et al. on the timing of mental events.

[7] Thanks to Amber Hopkins, Uri Maoz, Claire Simmons, and Walter Sinnott-Armstrong for comments on an earlier version.

Amber Hopkins

Gavenas, Hallett, and Maoz describe two possible roles for consciousness in action generation: generative or modulatory.[8] Would you agree more with the suggestion that consciousness is a generative source of action or that consciousness has a modulatory impact on action outcomes? You save a causally relevant place for unconscious aspects of conscious percepts in behavior. Would the conscious aspects have an impact on the unconscious aspects? Would this align more with the idea of consciousness as having a modulatory role in behavior?

David Silverstein and Hans Liljenström

Can unconscious decisions cause physical actions? For example, consider rapid and perhaps unconscious reactions by fear, such as a surprise encounter with a spider or snake (or something that appeared like one). Some hypotheses assert that signaling from threat-relevant stimuli can traverse a rapid and subcortical low road that activates the amygdala prior to conscious access and that amygdala efferents can trigger a response. Given this, can conscious awareness of an action and perhaps the awareness of fear emerge after the physical action?

Walter Sinnott-Armstrong

You argue forcefully that "the conscious aspect was causally efficacious" in your commonsense example of choosing a salad. However, it is still not clear *how* this causation works.[9] Gavenas, Hallett, and Maoz conclude, "[F]urther advances in the conscious control of movement are required to understand how those [conscious, subjective] experiences lead to action."[10] Do you agree with them that we do not yet know how any of this works? Can we be justified in believing that

---

[8] See Chapter 25 by Gavenas et al. on conscious control of action.
[9] See Chapter 26 by Hopkins et al. on consensus on consciousness and volition.
[10] See Chapter 25 by Gavenas et al. on conscious control of action.

the conscious aspect of our mental states do cause us to eat the salad when we do not know how one causes the other?

Replies to Follow-Up Questions

Ned Block

Color or mass?

When the heavy red brick breaks the window, the mass is causally efficacious, but the color is not. Mark Hallett and Liad Mudrik want to know whether the conscious part of a mental event that causes a behavioral response is like the redness of the brick or like the mass of the brick.

The issue my essay was concerned with was the allegation that because an unconscious part of a decision inevitably precedes all conscious aspects, it follows that the conscious aspects are like the redness of the brick rather than the mass of the brick. That is, the allegation is that the conscious aspects of the decision *never* are causally efficacious.

My response was to note that *every* phenomenally conscious mental event has an unconscious part that precedes all conscious parts. In conscious visual perceptions, unconscious activations of V1 precede the conscious aspects of the perception. But that does not show that the conscious aspect is like the redness of the brick. The way I argued for that was by pointing to the fact that in some cases, the conscious part and the unconscious part influence action in opposite ways. When red and green lights flicker at frequencies above 10 Hz, the subject sees yellow, but the unconscious processing activations in V1 and the lateral geniculate nucleus that precede the conscious experience of yellow register red and green. This opposition can give rise to contrary tendencies in behavior. If the subject presses the button for "yellow," we know that the conscious aspect of the perception was causally efficacious, that is, like the mass of the brick rather than the redness of the brick. If the subject presses the button for "red and green," we know that the unconscious aspect has dominated the response.

For all we know, cases in which conscious and unconscious aspects of a mental event push in opposite directions are a rare exception. Still, the existence of these cases is enough to show that the fact an unconscious part of a mental event inevitably precedes all conscious aspects leaves it open whether consciousness is causally efficacious. At least sometimes, consciousness is like the mass of the brick, not its color.

Generation or modulation?

Amber Hopkins asks whether consciousness has a generative or a modulatory effect on action. Let us ask this question with regard to the example of the perception of flicker that I mentioned earlier. We could say that the unconscious representations of red and green in the lateral geniculate nucleus and V1 have a generative effect but that (when the flicker rate is above 10 Hz) the conscious representation modulates that effect by turning the representation of red and green into a representation of yellow. But we could equally well say that the conscious representation of yellow has a generative effect on the percept since there is no unconscious representation of yellow. Or we could say that the unconscious processing generates the perception itself, whereas the conscious processing generates the content of the perception. These alternative ways of talking suggest that the generative/modulatory distinction is not as useful as one might have thought. If we are going to use the generative/modulatory distinction, we should take care to say whether we are talking about the mental event itself or its content. What generates the mental event might not be what generates its content.

Unconscious decisions?

David Silverstein and Hans Liljenström ask whether an unconscious decision can cause an action. What I would like to have available for an answer is a decision case that is parallel to the flicker and spatial frequency cases for perception, but there are two problems in finding such cases. The problems are illustrated by the example given by Silverstein and Liljenström in which exposure to a threatening stimulus might activate the amygdala via a fast, subcortical pathway, triggering a response prior to conscious perception of the threat. What they are considering is an *entirely* unconscious decision. The first problem is that the existence of entirely unconscious mental events is less certain than mental events that are partly conscious and partly unconscious. My perception examples were of the partially conscious/partially unconscious sort. When a perceiver sees red and green lights that flicker at 12 Hz, the resulting perception is partly conscious and partly unconscious. The conscious aspect registers yellow, whereas the unconscious part registers red and green. The reality of unconscious mental effects on action should not be taken to depend on whether the conscious part of a mental event can be "shaved off" to yield an entirely unconscious mental event.[11]

The second problem is that in the case of a putatively entirely unconscious decision, there is always an issue as to whether it is a decision at all as opposed to a sub-personal event—something more like a reflex. When the subcortical "low road" activates the amygdala, leading to the triggering of a fear response, is the triggering or some other part of this process really a decision at all?

Despite these problems, there are decision cases that share some important properties with the perception examples I described earlier, and I will now describe one. My example will not be totally parallel to the perception examples, since conscious and unconscious aspects of the decision process are woven together. What is parallel about the case, though, is that both conscious and unconscious aspects contribute to causing an action. Marc Jeannerod did an experiment involving three dowels that could be illuminated by computer-controlled diodes at their bases (Castiello, Paulignan, & Jeannerod, 1991). Subjects were asked to quickly grasp a dowel when it lit up and to lift the dowel up. In separate trials, subjects were asked to vocalize "Tah" when they saw the illumination and to grasp the lighted dowel. The start of reaching took 330 ms on average, whereas the start of the "Tah" sound took 380 ms on average. Unsurprisingly, when they did both tasks together, the hand movement started 50 ms before the vocalization.

Twenty percent of the time the illumination would shift between one dowel and another just as the reach started. Subjects were asked to say "Tah" again when they detected a change in illumination. In the

[11] See my contribution to Peters, Kentridge, Phillips, and Block (2017) for more on this point.

20% of cases where the light shifted, subjects started reaching toward one dowel and then, in 100 ms, course-corrected, grasping the second dowel. The experimenters also ran trials with the vocalization but no motor task and with the motor task but no vocalization. The main result is that the vocalizations occurred 315 ms after the subjects started to correct the trajectories of their motor movements despite the fact that control of the speech system is faster than control of the motor system. The time taken to produce a vocalization did not depend on whether or not there was also a motor movement and conversely, so it seems as if the two processes were substantially independent. Jeannerod argues from plausible assumptions about both motor processing and language processing that the course correction preceded awareness of the changing illumination and so must have been unconsciously generated. As Jeannerod (1997, p. 85) notes, subjects "reported that they saw the light jumping from the first to the second object near the end of their movement, just at the time they were about to take the object (sometimes even after they took it!)." Jeannerod told me that subjects reacted by saying their hand movement seemed to precede their awareness.

This case is not exactly parallel to the perception cases I mentioned. Both the original decision and the decision to course-correct occur in a conscious envelope, since subjects must have decided in advance to course-correct on the condition of seeing the light change. Still, the implementation of the original conditional decision is unconsciously generated, whereas the decision to grasp the original lighted dowel is conscious. So we do see both conscious and unconscious effects pointing in different directions in this example.

Whether or how?

Walter Sinnott-Armstrong suggests—and I agree—that we do not know how the conscious control of action works. But then, he asks, how can we be justified in supposing that there is any conscious control of action at all? A first level of reply would be to note that in the Jeannerod experiment, features of the course correction are unconsciously controlled, whereas the initial reach is consciously controlled. But how do we answer a skeptic who says the initial reach is really unconsciously controlled, too—given that we don't know how conscious control works? Such skepticism is undermined by the perception examples. We know that when the subject perceives the flickering red and green lights as yellow and presses the button for "yellow," there is conscious causation. But once we have allowed some conscious causation, why should we think that it never happens in the case of decision?

13

How is consciousness related to freedom of action or will?

Tim Bayne

Consciousness has been related to freedom of action and the will (hereafter "freedom") in a number of ways. Let's first explore a connection between consciousness and freedom that involves the experience of being free.

At your leisure, decide to raise either your left hand or your right hand, and then slowly raise it. Now, attend to the experiential aspects of this brief episode of agency. What was it like for you as an agent to decide between two possible intentions, and to then implement that intention? One influential answer to this question is that giving an account of the experiential aspect of ordinary human behavior requires that we recognize an "experience of freedom." Different theorists will provide different accounts of what exactly this experience involves (see, e.g., Bayne, 2008; Horgan, Tienson, & Graham, 2003), but roughly speaking we might gloss it as feeling that the particular action at which the experience was directed (e.g., the raising of one's arm) was "up to me."

There are at least two ways in which the experience of freedom might be related to freedom itself. First, it might explain why belief in the reality of free will is as widespread as it is. The thought here is that people tend to believe that they are free because they experience themselves as free. Second, one might appeal to the experience of freedom as providing evidence for the claim that we really do have free will. The fact that we experience many of our actions as "up to us" might not prove that they really are up to us (for these experiences might not be veridical), but—some have argued—if experience is generally a source of evidence, then we should regard experiences of freedom as *evidence*

Tim Bayne, *How is consciousness related to freedom of action or will?* In: *Free Will.* Edited by: Uri Maoz and Walter Sinnott-Armstrong, Oxford University Press. © Oxford University Press 2022.
DOI: 10.1093/oso/9780197572153.003.0013

for the reality of freedom. (As a parallel: the fact that I seem to see a cat on the mat doesn't prove that there is a cat on the mat, but it might provide evidence for that claim.)

Let's turn now to the idea that freedom requires consciousness of a certain kind. This idea is widely endorsed, but different theorists have very different conceptions of both the kind(s) of consciousness that are required for freedom and why consciousness is required for freedom.[1] Progress in this area requires making some distinctions.

First, we need to distinguish consciousness as a property of an agent ("creature consciousness") from consciousness as a property of mental states, such as an intention or a decision ("state consciousness"). Does freedom require creature consciousness, state consciousness, or both creature consciousness and state consciousness? In other words, does freedom require only that people themselves are conscious, or does it also require that some of their mental states—such as their intentions, beliefs, or actions—are conscious?

Let's begin with creature consciousness. If creature consciousness is understood in terms of *wakefulness* (as it sometimes is), then it seems likely that consciousness is necessary for freedom. After all, wakefulness is arguably necessary for the kind of rational integration and decision-making competence on which freedom depends, and sleepwalkers are not seen as acting freely. But creature consciousness is not always understood in terms of wakefulness. An alternative view holds that an agent is conscious in the sense of having *phenomenal consciousness* just in case there is "something that it is like" for the agent to be the agent that it is. This phrase is meant to capture the idea that being conscious is a matter of having an experience perspective—a subjective point of view. It is not obvious that freedom requires creature consciousness in this phenomenal sense of the phrase. After all, there doesn't seem to be any contradiction in supposing that my zombie twin—who has the same functional states that I do but has no experiential perspective—could have free will.

Of course, consciousness might be necessary for free will even if there is no contradiction in supposing that free will might obtain in

---

[1] See Chapter 7 by Hall & Vierkant on free will in degrees and Chapter 8 by Mele on free will.

the absence of consciousness. Suppose that the nature of consciousness is such that only creatures with an experiential point of view can have the kinds of deliberative and rational capacities that are needed for freedom. If that turns out to be the case—as, for all we know at present, it might—then free will would require creature consciousness in both of the senses we've identified: wakefulness and also phenomenal consciousness.

Thus far we have focused on consciousness as a property of agents. Let's now consider consciousness as a property of mental states. Does freedom require that certain kinds of mental states are conscious? If so, which ones—and why?

Different theorists will answer this question in different ways, but it is common to suggest that freedom with respect to a particular action requires being conscious of the decision and/or intention that initiates and/or structures it.[2] What exactly it means for a decision or intention to be conscious in the required sense is an interesting issue. It seems clear that being conscious of a decision or intention by (say) being told about it or inferring its existence from one's own behavior isn't the kind of consciousness that is required for free will. Instead, consciousness appears to require direct awareness of one's decisions and/or intentions.

Why might freedom require being directly aware of one's decisions and/or intentions? A number of proposals might be considered here. One answer is that one must be directly aware of one's decisions/intentions in order to have the kind of control over one's actions that is required for freedom. Another answer is that one must be directly aware of one's decisions/intentions in order to have immediate knowledge of what one is doing, and immediate knowledge of one's actions is required for freedom. A third answer is that one must be directly aware of one's decisions/intentions in order for those decisions and intentions to truly be said to be one's.

There is a great deal more that might be said about the ways in which consciousness might (or might not) be related to freedom of the will and action, but it should already be clear that answering this

---

[2] See Mudrik's and Schurger's replies in Chapter 24 on consciousness.

question raises some of the deepest challenges in our understanding of the mind.

Follow-Up Questions

Mark Hallett

The question here depends, as most questions, on definitions. However, if freedom of an agent is the ability to act without limitation or barrier, then it certainly can occur without consciousness. It would seem that the role of consciousness is just to inform the agent that it has freedom or that its acts are free. Right?

Walter Sinnott-Armstrong

As you say, understanding how consciousness is related to action and free will raises several of the "the deepest challenges" in philosophy of mind. Is part of what makes these issues so challenging "the absence of a consensus about the neural basis of consciousness and volition?"[3] What role(s), if any, could neuroscience play in meeting these challenges?

Adina Roskies

Can you say more about what it is to be directly aware of your intention? Certainly it is not that that intention be an occurrent object of thought, which is what it is to be conscious of, say, the cat on the mat. When I choose to step on the brake at a yellow light as opposed to choosing to speed up, I am not thinking "Now I am consciously choosing to step on the brake"—I am conscious of the red light and just make the choice. So the kind of awareness is not a meta-level awareness of the intention

---

[3] See Chapter 26 by Hopkins et al. on consensus on consciousness and volition.

itself. I think that many people have a mistaken view of what a conscious intention entails or is, parallel to the intuition of what it is to be conscious of an object. Do you agree?

Replies to Follow-Up Questions

Tim Bayne

Mark Hallett suggests that if freedom of an agent is the ability to act without limitation or barrier, then "it certainly can occur without consciousness." I'm not so sure. Why? Because giving an account of the nature of agents and of agency might turn out to involve consciousness. Consider the anarchic hand syndrome, in which a patient's anarchic hand might take food from the plate of another diner in a restaurant. Is this the action of an agent who is acting without limitation or barrier? Or does the absence of a certain type of consciousness mean that we have behavior but not an action, or action without an agent, or—perhaps—action and an agent but no free will?

Walter Sinnott-Armstrong asks whether the absence of a consensus about the neural basis of consciousness and volition might be part of what makes understanding the relationship between consciousness and free will so challenging. I tend to think not. I accept that findings in neuroscience could inform our understanding of how consciousness and free will are related. After all, *in principle* one's views about the demise of the dinosaurs can constrain one's views about the causes of the Russian Revolution. Nonetheless, *in practice* I suspect that understanding the neural basis of consciousness or volition is unlikely to shed much light on the relationship between these two domains, although it might help around the edges.[4]

Part of the issue turns on what exactly is meant by the "neural basis" of consciousness and volition. Consider a theory, *T*, that identifies the neural correlates of consciousness and volition but says nothing about *why* certain types of neural activity are associated with consciousness

---

[4] See Chapter 9 by Roskies on neuroscientific evidence about free will.

and/or volition whereas others aren't. I don't think that T is likely to contribute much to our understanding of the relationship between consciousness and free will. Now consider another theory, T^*, which not only identifies the neural correlates of consciousness and volition but also explains why certain neural regions and activity are centrally implicated in consciousness and/or agency while others aren't. T^* might well go some way toward clarifying the relationship between consciousness and agency. But is T^* likely to be a neuroscientific theory? I'm doubtful. Neuroscience might contribute to the development of T^*, but I doubt that neuroscience alone has the tools to develop T^*.

To my mind, understanding how consciousness is related to free will requires getting clear on the contribution(s) of consciousness to a range of high-level processes: decision-making, deliberation, rationality, inhibition, perspective-taking, and many others. Neuroscience can certainly tell us much about the implementation of these capacities, but it is psychology that is best placed to reveal their nature.

Adina Roskies asks me to say more about what it is to be directly aware of one's intention. I regard the notion of a conscious intention as a many-headed beast, and I suspect that many of the debates in this area occur because people focus on different conceptions of a conscious intention (unaware, perhaps, that other senses are viable).

Two notions strike me as central. In what we might call the *deflationary* sense, an intention is conscious in virtue of the fact that it is part of one's overall subjective experience. Typically, intentions are conscious in this sense because they structure one's agency, and one is aware—if only in a background kind of way—of what one is doing. In walking I might have a kind of background awareness of my intentions to move my legs, for I am aware that I am walking, and insofar as I am walking (that is, acting in a certain way) then I will also have intentions that govern my movements. In what we might call the *inflationary* sense, an intention is conscious if it is the object of explicit attention. Perhaps the clearest examples of inflationary conscious intentions occur after deliberation, when one decides on a certain course of action. Taxonomy is tricky here, but I tend to think of deflationary intentions as (or akin to) perceptual experiences and inflationary intentions as (or akin to) thoughts.

Do intentions of either kind occur when one steps on the brake in response to a traffic light turning amber? Inflationary intentions certainly don't, and perhaps talk of choice and decision-making is also out of place here. I'm less sure what to say about deflationary intentions, but I am inclined to think that insofar as one experiences braking as something that one is doing (as opposed to a mere movement of one's body), then there is a role for a conscious intention here, albeit one that occupies only the margins of awareness.

QUESTIONS ABOUT RESPONSIBILITY AND REASONS-RESPONSIVENESS

14

How is responsibility related to free will, control, and action?

Gideon Yaffe

A plausible quartet of philosophical hypotheses is that for a person to be responsible for her behavior it must be (1) action—that is, something she *does* rather than something that merely befalls her—which is (2) under her control, (3) freely chosen, and (4) freely performed. A person who lacks either freedom of action or will, or who is not in control when she acts a certain way, is not responsible. Similarly, if the behavior is not action, then, too, there can be no responsibility for it. However, all four claims have been challenged.

A wellspring of reflection on this topic is an example from John Locke (1689/1975): a man is brought into a room while asleep, and the door is locked. He wakes and finds a friend in the room, so he decides to stay to visit with her. Is he responsible for staying? It appears so. If he missed an important meeting thanks to having stayed, then those who were depending on him would have grounds for complaint. But the man could not have left the room, since there was a lock on the door. This fact was unknown to him and so played no role in the psychological process that led him to decide to stay. But, still, the lock took away the man's ability to do otherwise than stay, and yet it did not take away his responsibility for staying. Thus, it appears, freedom of action, which requires the ability to act otherwise, is not necessary for responsibility after all.

It is natural to respond that the man could have *tried* to leave the room. Perhaps he is responsible not for staying, simpliciter, but for staying *willingly*. Had he struggled to open the door in an effort to get out, something he was capable of doing, then he still would have stayed, but he would not have stayed *willingly*. So the example does not

Gideon Yaffe, *How is responsibility related to free will, control, and action?* In: *Free Will*. Edited by: Uri Maoz and Walter Sinnott-Armstrong, Oxford University Press. © Oxford University Press 2022. DOI: 10.1093/oso/9780197572153.003.0014

threaten the hypothesis that freedom *of will* is necessary for responsibility, only that freedom *of action* is.

However, examples concocted by Frankfurt (1969), and now referred to as "Frankfurt examples," suggest that Locke's example can be extended to show that there can be responsibility without freedom of will, as well. Frankfurt imagined that a person has a device installed in his brain that will cause him to *choose* to do something if a button is pressed. A diabolical person waits to see whether the person will make the choice on his own, planning to press the button if he will not. As it happens, the person does make the choice on his own—imagine it to be a choice, for instance, to steal money—so the button never needs to be pressed. This person is responsible for stealing the money. But he did not act with free will when he stole, for had he been about to choose not to steal, the button would have been pressed, and he would have gone on to steal. This person could not even *try* not to steal, for the effort would require a choice not to steal, and had he been about to make such a choice, the button would have been pressed before he could choose not to steal. Essentially, Frankfurt moved the lock on the door in Locke's example into the head. The implanted device serves as a factor that ensures a certain choice or will but does not actually cause it since the agent generates the choice through an independent mechanism, the involvement of which suffices for responsibility.

Some philosophers, known as "semi-compatibilists" (a term coined by John Fischer, 1988), have responded to this dialectic by asserting that the abilities to choose and act otherwise are necessary for the kind of control needed for freedom of will and action but are not necessary for the kind of control required for responsibility. This position leaves open the question of what, exactly, it is to be in control of choice or conduct in the sense required for responsibility. To explain this notion, much philosophical work has focused on human capacities to recognize and respond to reasons for action.[1] The man in the locked room recognizes a reason to stay (it would be nice to spend time with the friend); he neglects a reason to leave (others are depending on him); and this psychological mindset is what leads him to stay in the room. He thereby demonstrates a form of control that is missing in, say, the bodily motions involved in

---

[1] See Chapter 15 by Sinnott-Armstrong on reasons.

an epileptic seizure, which are in no way guided by the agent's recognition of reasons. Semi-compatibilists have held that it is this kind of control—guidance of choice and conduct by reasons—that is necessary for responsibility, and not the kind of control that is enjoyed by those who have the ability to choose or do otherwise than they do.

When an agent's bodily motions are guided by her recognition of her reasons, they are things *done* by the agent, not things that happen to her, like heartbeats or convulsive motions. It can therefore appear that, if what is necessary for responsibility is the kind of control we exercise when we are guided by reasons, then action, too, is necessary for responsibility; whenever we exercise that form of control, we act.

However, this claim, too, has been questioned. Consider someone who forgets her spouse's birthday. This is something for which she might be responsible, but is it *action*? Sometimes forgetting is caused by action or omission for which we are responsible, like omitting to tie a string around one's finger. But does the injured spouse need to be able to identify something that the forgetful partner failed to do and should have done in order to hold her responsible? The consensus view is that in cases like this there can be responsibility without action (Murray, Murray, Stewart, Sinnott-Armstrong, & De Brigard, 2019). This in turn prompts the question of whether every manifestation of an agent's dispositions for recognizing and responding to reasons is an action. Those who think not are able to hold on to the idea that what is necessary for responsibility is neither freedom of will nor freedom of action but, instead, is guidance by reasons. Those who think so, by contrast, must deny that even responsiveness to reasons is necessary for responsibility.

So, while there is not a settled view among philosophers as to whether freedom of action and will, control, and action are each necessary for responsibility, there are strong reasons to doubt the simple, unqualified assertion of any of these four claims.

Follow-Up Questions

Uri Maoz

Your chapter focuses on a conceptual analysis of the relations between responsibility, free will, control, and action. However, do you think that some parts of this debate are empirically tractable? Can neuroscience play a role in this debate? If so, which parts and role?

Deniz Arıtürk

You discuss the proposition that responsibility requires guidance by reasons. Does this proposition apply to all reasons or only to those that arise out of specific sources, such as the agent's "deep self"? For instance, is the addict's strong desire to drink an appropriate kind of reason to hold her morally responsible for her choice to drink? And is the delusional patient's desire to save her own life the appropriate kind of reason to allow us to hold her morally responsible for killing someone whom she believes is trying to kill her?

Liad Mudrik

Can agents be responsible for arbitrary actions? The different definitions and criteria for responsibility reviewed in this chapter seem to provide different answers to this question. Is there any consensus about the status of arbitrary decisions with respect to responsibility?[2]

Reply to Follow-Up Questions

Gideon Yaffe

The promise and limits of the neuroscience of free will
It is perilous to specify limits about what a science can and cannot help us to understand. Discoveries in every branch of science frequently come to bear on questions that had seemed, in advance, beyond the scope of that branch of science to answer. Still, it appears that with its

---

[2] See Chapter 19 by Bold et al. on arbitrary and deliberate decisions.

current tools, neuroscience can, potentially, be useful for answering some questions about free will and responsibility, and not others.

One set of questions about responsibility, action, control, and freedom concerns *criteria*: What are the necessary and sufficient conditions that an event must meet for a person to be responsible for that event, or for it to be action, or for it to be under her control, or for it to have been chosen or performed freely by her? The answers to these questions are not merely stipulative or verbal; we are not agreeing on how to use a word. Rather, competing answers are assessed by their comparative capacity to explain and systematize our practices. An account of the criteria for action, for instance, has something to be said for it to the extent that it explains, among other things, what we put in and exclude from the category, what inferences we are willing to make about actions and not about events that are not actions, what judgments and emotional responses we have toward the actions of ourselves and others that we do not have with respect to events that are not actions, and so on. While the behavioral sciences, and neuroscience, can help us to understand what role the concept of free and responsible action plays in our ordinary lives, it can seem that neuroscience is powerless to provide us with an account of the criteria that an event must meet for it to explain and systematize these experiences. Of course, we might find out things about the brain that help us to determine whether or not people *meet* the relevant criteria—and that's important—but it is tempting to say that no measurements of the brain can help us to determine what the criteria themselves are.[3]

While there is much to this idea, it is slightly too simple; there is at least one way in which neuroscience might help us to specify the relevant criteria. The reason is that the philosophically optimal answer to the criterial question might involve deference to the discoveries of science. For instance, philosophers concerned with the criteria for action have tried to solve what is sometimes called "the problem of deviant causation." A famous example of Donald Davidson's illustrates the problem: A and B are climbing a cliff together. A intends to kill B, and he has an opportunity for he is holding a rope that is preventing B from falling to his death. A thinks to himself, with alarm, "I now have the

---

[3] See Chapter 9 by Roskies on neuroscientific evidence about free will.

opportunity to do as I intend and kill B!" This thought makes A very nervous, so nervous that his hands shake, and thanks to the shaking he lets go of the rope, causing B to fall and die. A's hand motion was caused by A's intention, and they caused B's death. But A's hand motion was not action; the motion was caused "deviantly" by A's intention. But what is the difference between deviant and non-deviant causation? One possible answer is that non-deviant causation is the *normal* kind, the kind that we find in normal cases of intention execution. And, perhaps, neuroscience can help us to determine what happens in the normal case. Neuroscience, then, would help us to answer the criterial question since the criteria themselves involve deference to the discoveries of neuroscience.

In addition, as already indicated, neuroscience can help us to understand the *extent* to which we satisfy the criteria for action or freedom. For instance, it has sometimes been thought that human beings have the capacity to act for a particular reason and to the neglect of others who favor the same course of action. You like the actress Emma Stone, and you like musicals, but you decide to watch the film *La La Land* because it has Emma Stone in it and not because it is a musical. It often feels, in cases like this, as though you could have, instead, watched the film for the neglected reason, in this case because it was a musical. But whether this is so is, in part, an empirical question that does not seem out of the scope of neuroscience, or perhaps neuroscience conjoined with psychology, to answer. The question is only an empirical question "in part" because before you can answer it empirically you need criteria for the ability to choose otherwise. The discovery of such criteria is in the first instance a philosophical undertaking. But, still, once you have such criteria, neuroscience could help us to determine if we meet them.

Further, it is also possible that neuroscientific discoveries about the degree to which human beings satisfy independently specified criteria for action, control, freedom, or responsibility will in turn influence our philosophical accounts of those criteria. The reason is that part of what a good account of the criteria will do is to align, more or less, with our judgments about the extent to which they are realized. If neuroscience were to teach us that no human being meets proposed criteria for action, that might make us doubt whether those criteria are satisfactory.

Isn't it evident, after all, that people sometimes act? So this is yet another way in which neuroscientists and philosophers can work together to get a better picture of free and responsible agency.

Guidance by reasons

While responsibility requires guidance by reasons, such guidance is insufficient. If the notion of "guidance" is specified with care, so that causation of a behavior by a reason is insufficient for guidance (as deviant causation cases show), then guidance by reasons may be sufficient for action or even for intentional action. But even intentional action is insufficient for full responsibility. Examples of alienation from one's own predilections to treat a fact as reason-giving illustrate the point. The addict treats the fact that by stopping at the bar he can spend the night drunk as a powerful reason to stop at the bar, a reason that is to him in the moment much stronger than the reason to pass by the bar and go, instead, to his child's piano recital. When he stops at the bar, he does so intentionally. But if he is powerfully alienated from the facts about himself that lead him to treat the prospect of getting drunk as so strongly reason-giving, then that can lead us to view him as at least diminished in responsibility for his intentional actions. He is, in any event, less responsible for the harm that he inflicts than is a non-addict who does the same thing or than an addict who is in no way alienated from himself. These facts matter to responsibility.

Cases of truly arbitrary behavior, performed for no reason at all, are not behaviors for which we are responsible. But they are also not what people ordinarily mean when they talk about "actions performed for no reason." Virtually every action that we are tempted to describe that way was in fact performed for a reason under an alternative description. Why did you choose that jar of jam from the grocery store shelf, rather than another, identical jar? No reason. But there is a reason why you chose *a* jar of jam; you were planning to spread it on toast. The appearance that the act is performed for no reason derives from the question that we asked about it; we sought a "rather-than" explanation, and there wasn't one handy. If the act is one for which we want to hold you responsible—say that, for instance, the jam company has terrible labor practices for which we hold its loyal customers in part accountable—it will be in part because of the fact that you bought the jam for a reason, and to the neglect of powerful reasons not to. When we think of examples of behaviors that are not performed for reasons under any description that applies to them, they stop appearing to be actions at all; they are merely things that happen to us. But it is not the case that things that happen to us are things for which we are responsible; action is necessary for responsibility. So truly arbitrary actions are not things for which we can be responsible, but they are also rarely those about which the question of responsibility is raised.

15

What are reasons?

Walter Sinnott-Armstrong

Why do you believe that? Why should I do that? Why did that happen? These why-questions ask for reasons of different kinds. Their answers give, respectively, *epistemic* reasons to believe some proposition, *practical* reasons to carry out some action, or *explanatory* reasons why some event occurred. It is important not to confuse these kinds of reasons.

The question "Why did you do that act?" could ask for either of the last two kinds of reasons. Questioners are asking for *explanatory* reasons when they want you to explain how you could have done something bad, such as cheat on your spouse or your taxes. Your explanation might cite your desire or motivation ("I was in love" or "I was desperate for money") and perhaps also a belief (such as that you would get away with cheating). Such explanatory reasons need not show that the act was the right thing to do.

In contrast, people who ask "Why should I do that?" want to know what makes that act right, such as what makes them justified in revealing a certain secret. Then they are asking for *practical* or *justifying* reasons that make the act normatively proper in some way. Some practical reasons might show why the act is in the agent's self-interest, so it is *prudentially* right or rational, but other reasons might show why the act is *morally* right in an impartial way even if it is against the agent's self-interest. Prudential reasons and moral reasons are only two kinds of practical reasons.

In addition, people sometimes ask "Why did you believe that this act was right?," such as "Why did you think that sacrificing your bishop would help you win our game of chess?" People who ask this question might already know your motive (to win) and also know whether the sacrifice was good or bad (depending on whether you won or lost), so

Walter Sinnott-Armstrong, *What are reasons?* In: *Free Will*. Edited by: Uri Maoz and Walter Sinnott-Armstrong, Oxford University Press. © Oxford University Press 2022.
DOI: 10.1093/oso/9780197572153.003.0015

they are not asking for explanatory or practical reasons. Instead, they are asking for a reason that is *epistemic* in the sense that it is a reason for belief instead of action and it is about evidence for that belief that could make you justified in believing it. If your bishop sacrifice did work, they might want to understand how you knew it would work. And if it did not work, they might want to understand what misled you. Either way, their question is "What were you thinking?"

Notice that answers to why-questions about epistemic reasons are contents of *beliefs* of people who have the epistemic reasons. I made that chess move because I believed that my opponent would make this move next. In contrast, answers to why-questions about explanatory reasons are *desires* or motivations of the agent. I cheated on my taxes because I needed money. In contrast, answers to why-questions about practical reasons are *facts* about the act that the reason justifies. What made it right to reveal the secret is the fact that a judge ordered me as a witness to answer the attorney's question.

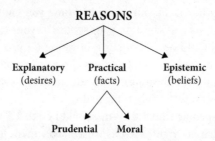

Puzzles about reasons arise when we conflate different kinds of reasons. For example, philosophers disagree about whether reasons are causes. A solution is to realize that desires or motivations that are explanatory reasons why an agent did an act are causes of that act. In contrast, practical reasons that justify that act and make it the right thing to do are not causes of the act, because they are facts about the act or its consequences. The *fact* that buying a security system will reduce burglaries of my home is a (practical) reason why I ought to buy a security system, but my subjective *belief* in that fact along with my *desire* to prevent burglaries of my home are what cause me to buy a security system and provide (explanatory) reasons why I bought one.[1]

Similarly, when neuroscientists want to investigate the neural correlates of reasons, we need to determine whether they are asking about beliefs that are epistemic reasons, desires that are explanatory reasons, or facts about the act that are practical reasons. The first two reasons are mental states—beliefs and desires—that are reducible to or emerge from brain states.[2] In contrast, a practical reason can be a fact that is not about the brain states of the agent. The biological fact that lilies are poisonous to cats is a reason for me to stop my cat from eating lilies, but that fact about cats is not about my brain, even if I need a brain in order to recognize that fact. Practical reasons like this biological fact about cats do not have neural correlates, even if epistemic and explanatory reasons do.

Despite these differences, we sometimes refer to all three kinds of reasons at once. Suppose you ask me, "Why did you bet so much on the game?," and I answer, "Because I knew I would win a lot of money." This single answer refers to a practical reason that justifies my bet (the fact that I was going to win a lot), an explanatory reason (my motivation to win a lot of money), and an epistemic reason (my belief or evidence that I would win). Although this one sentence gives all three kinds of reasons at once, these three kinds are still distinct, because sometimes one can be present without the others.

Philosophers disagree about what is common to these three kinds that makes them all reasons. My view is that they all play a certain role in reasoning (Hieronymi, forthcoming) or argument (Sinnott-Armstrong, 2018). An argument for a certain conclusion can either justify belief in that conclusion (by presenting an epistemic reason) or explain why that conclusion is true (by giving an explanatory reason). When the conclusion is about what ought to be done, an argument can also give a practical reason to justify the act in the conclusion. In all of

---

[1] See Chapter 1 by Yaffe on intention.
[2] See Chapter 23 by Hopkins & Maoz on the neural correlates of beliefs and desires.

these cases, the premises of the argument are reasons because of their function in the argument.

Despite this relation to reasoning and argument, a person can have a reason to believe or do something without that person explicitly or consciously formulating any chain of reasoning or any premises in an argument. I can quickly respond to my reason to run away from a nearby hissing snake without explicitly thinking "This is a snake; snakes are dangerous; I should run away from dangerous things; therefore, I should run away from this snake." Reasons do not have to work step by step in the way that explicit and conscious reasoning and argument do, even though what makes them reasons is their potential role in a structure of reasoning or argument. As a result, it can sometimes be difficult for people themselves (as well as others who observe them) to know which reasons they are responding to, if any.

Reasons also do not have to be overriding. I can have *some* reason to play golf (I know my buddies want me to join them) without having *overall* reason to play golf if I have overriding reason not to play golf (I have to go to work). When people say simply that there is reason to do something, it is often not clear whether they mean that there is an overall or overriding reason or only that there is some reason, which might be overridden by conflicting reasons.

A belief or act is *irrational* when there is an overriding reason against it, at least when the believer or agent is or should be aware of that reason. The mere presence of some reason not to believe a proposition or not to do an act is not enough to make the belief or act irrational when that reason is overridden by other reasons. The act or belief can still be rationally *permitted* (that is, rational in the weak sense of not being irrational) when there is some reason for it and no overriding reason against it, but it still might not be rationally *required* (that is, rational in the strong sense that not believing or doing it is irrational). For example, if I decide between two indistinguishable dollar bills, I am rationally permitted to pick the left one (that choice is not irrational), but I am not rationally required to pick the left one (because I am also rationally permitted to pick the other one). We need to distinguish these senses of rationality (rationally permitted versus rationally required) just as we need to separate strength of reasons (overriding versus overridden reasons) and kinds of reasons (practical, explanatory, and epistemic) in order to avoid confusion.

Follow-Up Questions

Mark Hallett

A good argument is made here for many types of reasons, and these are likely instantiated in the nervous system with different processes, but they would all likely appear to influence any final action or decision. Any action or decision would be the result of many causes, and these causes can be called reasons. Is there any situation where a reason (known to the agent) would not be a cause?

Gabriel Kreiman

Why did the coin land on head instead of tail? It does not sound too exciting to ask this question. We know about the process of free fall, we understand how rigid objects bounce, and we understand angular momentum. Period.

Why did you raise your right hand? One day we will understand the neuronal mechanisms that lead to raising your right hand, and the question will be as mysterious and as interesting as why the coin chose heads. The "reasons" should ultimately be described in the language of action potentials.

To my knowledge, humans do not have any reasons to raise their right hands any more than coins have reasons to choose heads over tails. A human could presumably have chosen to raise her left hand under *different* conditions, in a different scenario. (Your brain at 3:00 p.m. is not the same as your brain at 3:01 p.m.!) Given *identical* starting conditions at the neuronal level, however, the human would always have chosen the right hand. (This is a hypothesis, written as a true statement for emphasis.) The coin chose heads instead of tails because of a combination of how it was launched, its weight, its imperfections, the wind, the properties of the floor, plus a myriad of other factors.

There are probably a few tens of relevant variables involved. (Believe it or not, physicists have modeled this phenomenon!) For humans, there are probably tens of thousands of relevant variables involved in deciding to raise the right hand. Add the chemistry of carbon, mix in a few neurons, combine it with many more variables and the key ingredient of anthropomorphism, and we get "reasons."

Liad Mudrik

This question entails an assumption that there are specific reasons for each action that can be identified; for the sake of argument, let's accept that assumption (though I would love to learn if it is philosophically accepted). If so, what are the tools we have for identifying these reasons? Is asking the agent enough? Clearly not, as we acknowledge cases wherein the agent is not aware of the reasons that drove her to act. How can we identify the reasons for an action, then? And is there any criterion for determining that we identified them correctly?

Replies to Follow-Up Questions

Walter Sinnott-Armstrong

Thanks for these stimulating questions. They provide an opportunity for me to elaborate on some aspects of my brief account of reasons.

Are all reasons causes?
The first two issues concern relations between reasons and causes. Hallett asks, "Is there any situation where a reason (known to the agent) would not be a cause?" He refers to "action or decision," so he is not asking about epistemic reasons for belief, but his question could be about either practical or explanatory reasons. My answer differs for these two kinds of reasons.

Suppose there is a practical reason why I ought to get a vaccination, and I know that reason, which is the fact that the vaccination will protect me against a rampant disease. Nonetheless, what causes me to get the vaccination might be that my partner insists that I get it, and I might never have gotten it without my partner insisting. Then the practical *reason* I know is not the *cause* of my action.

In contrast, the reason that explains why I got the vaccination is my desire to appease my partner. That desire is a known explanatory reason and does cause my action. So my answer to Hallett is that explanatory reasons are causes, but practical reasons need not be.

Are all causes reasons?

Hallett also mentions the converse: "Any action or decision would be the result of many causes, and these causes can be called reasons." This claim assumes that all causes of actions or decisions are reasons. That assumption might hold for all *explanatory* reasons, if what causes my decision also explains it. However, it fails to hold for all *practical* reasons. If I can pick either of two cars that do not differ in any relevant way, then I can pick car 1, even if I have no practical reason to pick car 1 instead of car 2. What causes me to pick car 1 might be only a wind that blows me slightly closer to it, but that cause is not a practical reason to pick car 1 instead of car 2.

The same goes for neural causes. Kreiman writes, "The 'reasons' should ultimately be described in the language of action potentials." If I pick car 1 because the wind blew me closer to that car, that external cause makes neurons fire inside my brain, which in turn cause me to pick car 1. However, this internal, neural cause is still not a practical reason to pick car 1.

Why not? As I said, what makes reasons reasons is "that they all play a certain role in reasoning (Hieronymi, forthcoming) or argument (Sinnott-Armstrong, 2018)." The language of action potentials does not fit into an argument with premises that could *justify* the practical conclusion that I should pick car 1. "These neurons fired, so I ought to pick car 1" makes no sense as an argument. That is why descriptions at these levels do not provide *practical* reasons. They still might *explain* why I picked car 1 because they do fit into a different argument: "These neurons fired, so I did pick car 1." The fact that the language of action potentials provides *explanatory* reasons (in this argument) without *practical* reasons (in the previous argument) shows why we need to distinguish these different kinds of reasons.

Do humans have reasons?

Kreiman might reply, "[H]umans do not have any reasons to raise their right hands any more than coins have reasons to choose heads over tails." He is not denying explanatory reasons, for he predicts that scientists will be able to explain movements of both coins and humans. So he seems to be denying that humans have practical reasons.

Why? Kreiman argues, "Given *identical* starting conditions at the neuronal level, however, the human would always have chosen the right hand" (assuming the same external influences). I agree that human actions are determined, but determinism does not rule out practical reasons of any sort I would countenance.[3] Moreover, a coin does not have any desires, goals, or values. Since the point of practical reasons is to guide us toward our goals and values and thereby justify choices and actions, these differences between coins and humans suggest that humans can have practical reasons even if coins do not.

How can we know our reasons?

Even if humans do have reasons for actions, how can we identify *which* reasons they have? That is Mudrik's question. If she is asking about *practical* reasons, then an answer would need to apply normative theories about what we should do (prudentially or morally). Instead, she seems to have in mind *explanatory* reasons for actions, so I will focus here on those.

We often do not know many aspects of the reasons that explain why we act as we do. Imagine that I am hungry, so I eat some chocolate quickly. I identify my hunger as a reason to eat something, and my love of chocolate as a reason to eat chocolate. I still might not recognize any reason to eat quickly. Something *caused* me to eat quickly, but I might not be able to identify any *reason* or explanation for that aspect of what I did. Nonetheless, I do seem to know my reasons for some aspects of my action.

How can I know whether I am correct? As Mudrik says, asking an agent is not always enough to justify a belief that the agent acts for the reported reason. Nonetheless, if you ask me why I ate chocolate, and

---

[3] See Chapter 6 by O'Connor on determinism.

I reply, "I love it," this report is some evidence that my desire for chocolate was part of the explanation. This evidence is not conclusive, because I might lie or order chocolate only out of habit without really liking it. But evidence is often not conclusive.

If we want stronger and more precise evidence, we need to look at broader patterns. Do I always order chocolate, at least when I cannot instead get ginger, which I love more? Am I willing to pay more for chocolate when caramel is on sale? Evidence from neuroscience also might show whether or not we react to certain information that we are not able to report.[4] These and other kinds of evidence can make me more justified in believing that I identified the right explanation for my choice. None of this comes near to conclusive proof, but sometimes we have enough evidence for some degree of confidence about our explanatory reasons for some aspects of our actions.

---

[4] See Chapter 24 by Mudrik & Schurger on consciousness.

PART II

QUESTIONS
FROM PHILOSOPHERS
FOR NEUROSCIENTISTS

PART II, SECTION I
QUESTIONS ABOUT WILL

16

What are the main stages in the neural processes that produce actions?

Patrick Haggard and Elisabeth Parés-Pujolràs

In order to answer this question, we will consider that actions necessarily involve body movements, caused by muscle contractions. Such body movements are physical events, that can be recorded and measured as a matter of objective fact, but what are the causes of these events? We will try to answer this question in terms of motor neurophysiology, but we first point out that this question is also in a deep sense *psychological*. Psychology typically seeks to explain the causes of behavior; indeed, the defining question of psychology is "Why did she do that?" A description of the neural processes that lead to a body movement should ultimately include descriptions of the neural centers that plan, deliberate, and motivate future actions and that give our actions the role and value they have in our everyday life. So it is neurophysiology, with its established concepts of pathway and hierarchy, that offers a clear and rigorous way to answer psychological questions.

The neurophysiological way to answer this question involves reverse-engineering the motor pathway.[1] This means beginning with the overt physical movement of a body part and investigating the series of neural events that provided the causal antecedents of that movement. The body part moves because a muscle contracts. The muscle contracts because of drive from the spinal motoneurons. We can continue along the chain: Why do the spinal motoneurons fire action potentials? Here we find a first key difference that is clearly relevant

---

[1] See also Chapter 20 by Hallett on levels and control in the brain.

Patrick Haggard and Elisabeth Parés-Pujolràs, *What are the main stages in the neural processes that produce actions?* In: *Free Will*. Edited by: Uri Maoz and Walter Sinnott-Armstrong, Oxford University Press.
© Oxford University Press 2022. DOI: 10.1093/oso/9780197572153.003.0016

to volition. Three fundamentally different classes of neurons synapse onto the motoneurons in the ventral spinal cord. First, direct afferent signals from the muscle itself provide a simple reflex regulation of muscle activity. In particular, when a muscle is passively stretched, it rapidly and reflexively contracts, until it returns to its original length. An interesting example comes from catching a small, fast, and heavy ball, as people frequently do in baseball or in cricket. When the ball first hits the fielder's hand, it first stretches the muscles of the forearm. The stretch is signaled by afferent neurons to the spinal cord, which then synapse onto motoneurons that contract the same muscles, bringing the arm back with a rebound toward its original position. A second synaptic input to the motoneurons comes from the many interneurons in the spinal cord, which allow the reflex contraction to be modified and controlled.

However, from our point of view, the most important input to the motoneurons is the third input. This is the descending input from neurons in the cortex, and also from higher spinal areas and brain-stem centers. It consists of axons traveling in the corticospinal tract, propriospinal tract, and other white matter tracts. These inputs provide a "motor command" that descends in the spinal cord to instruct the muscle to contract, which typically results in the limb moving to a new position. The corticospinal tract is perhaps the best studied and most intriguing of these pathways, because it has traditionally been considered the "final common path" for voluntary action (Sherrington, 1952). This view has lasted well over the past century or so: voluntary actions typically involve firing of neurons with cell bodies in primary motor cortex (and perhaps also in adjacent areas) that send axons to the spinal cord on the opposite side of the body. In some cases, other descending tracts can substitute, but the motor cortex corticospinal tract is normally considered the output path par excellence for voluntary movement. The question of *how* neurons in the motor cortex compute and transmit the appropriate signals to muscles has been widely considered in the motor control literature.

For the current purpose of reverse-engineering a human action, however, a more relevant question may be *why* the motor cortex transmits a voluntary motor command. Several theories view the frontal lobes as implementing a processing hierarchy that progressively transforms abstracts plans and intentions (thoughts?)[2] into specific motor actions (Koechlin, Ody, & Kouneiher, 2003). In keeping with this view, a key input into the primary motor cortex comes from the cortical areas immediately anterior to it, essentially Brodmann's Area 6 (BA6).[3] Neurons in this area can be separated into two separate functional circuits. More lateral parts of BA6 receive extensive input from superior and inferior parietal cortex and are thought to link specific stimuli to the appropriate motor response. In contrast, more medial parts of BA6 (which also project corticocortically to primary motor cortex, and to some extent directly to the spinal cord as well) receive input from the prefrontal cortices, particularly the limbic regions, but remarkably little parietal input (Caminiti et al., 2017). This basic difference in anatomical connectivity seems highly relevant for "free will," as noted by both neurologists (Goldberg, 1985) and neurospsychologists (Passingham, 1987) some decades ago. In essence, when humans and other animals respond to demands or conventions of external stimuli, the lateral route is in play, but when they act based on an internal need, drive, or perhaps an intention, or for some other reason,[4] the medial frontal cortex is key.

The idea that medial frontal cortex plays a special role in voluntary action is borne out by many studies. First, the readiness potential (Kornhuber & Deecke, 1965), long held to be the classic biomarker of volition (Kranick & Hallett, 2013; but see Schurger, Sitt, & Dehaene, 2012), was thought to arise from these areas. This may have motivated the implausible proposal from the Nobel Laureate Sir John Eccles (Beck & Eccles, 1992) that "free will" involved a dualist interaction by which mental processes could intervene on variable physiological activity in the supplementary motor area. Second, stimulating in this area sometimes produces an experience of being about to move. Very interestingly, lesions to or reduced activation of the medial frontal cortex leads to two broad classes of neurological syndromes. The first is a poverty of voluntary action, as found in neurological syndromes

[2] See Chapter 1 by Yaffe on intention.

[3] BA6 is part of the motor cortex, comprising the premotor cortex and supplementary motor area. The premotor cortex and supplementary motor area are featured in Figures 2b and 3 in the Brain Maps section.

[4] See Chapter 15 by Sinnott-Armstrong on reasons.

of akinetic mutism or in apathy of some patients with basal ganglia disease (Muhammed, Manohar, & Husain, 2015). The second, in contrast, is a hyperkinetic disorder following medial frontal damage, in which the patient's actions are frequently and involuntarily captured by stimuli in their immediate environment. Thus, a patient with anarchic hand syndrome or utilization behavior might grasp objects that happen to be present and compulsively use them, even if they have no wish or reason to do so. This finding suggests that the healthy medial frontal cortex houses a mechanism that normally inhibits actions afforded by environmental stimuli and transmitted via lateral cortical routes. This inhibitory mechanism is presumably suspended when such affordances correspond to our current intention. One might postulate that medial frontal cortex houses two distinct processes. The first might be called "positive volition" and corresponds approximately to the philosophical concept of will.[5] It is the capacity to initiate endogenously those actions that correspond to our intentions. The second process might be called "negative volition" and refers to the capacity to withhold actions that are possible, or even tempting, but do not correspond to current intention. Negative volition is related to the social psychological concept of willpower (Baumeister & Vohs, 2003), and its deficiency might be related to the philosophical concept of akrasia.

Neuroscientists have a useful biomarker for the transition of action information anterio-posteriorly from the medial frontal cortex to the primary motor cortex. The latter is strongly contralateralized, such that the right hand is controlled predominantly by the left motor cortex, and the left hand is controlled predominantly by the right motor cortex. In contrast, medial frontal areas such as the pre-supplementary motor area[6] are much less strongly contralateralized. In EEG scalp recordings of neural activity prior to voluntary action, there is an identifiable shift from an early bilateral readiness potential to a later phase of lateralized readiness potential (LRP), contralateral to the hand that will move. The onset of the LRP often occurs

---

[5] See Chapter 2 by Hieronymi on the will.
[6] The pre-supplementary motor area is the anterior part of supplementary motor area (SMA). The SMA is featured in Figure 2b in the Brain Maps section.

some 500 ms prior to voluntary action. This neurophysiological event has often been used to provide insight into the processes of deciding what to do and when to do it: at the point that the LRP begins, the brain appears to have transitioned from a general intention to move to a specific intention regarding what body part to move (Haggard & Eimer, 1999; but see also Schlegel et al., 2013). By this point, a mere intention seems to have become a concrete plan, with means as well as ends.

One might continue the reverse-engineering project still further: What is the source of the neural signals that lead to medial frontal cortex activity? Clearly, the reverse-engineering project runs the danger of regressing to a "first cause" neural process that initiates voluntary actions but is not itself caused in the same way. Cartesian dualism offered an explanation of this form, but scientific accounts of voluntary action have to avoid this kind of causal exceptionalism. Ultimately, the most abstract and "intentional" areas of the prefrontal and frontal cortex must receive signals from somewhere, and the regress cannot simply be pushed further back toward a metaphysical placeholder "will." Here, neuroanatomy again offers useful clarity: the key areas for voluntary action in the prefrontal and medial frontal cortex are all also targets of the subcortical loops that run from the cortex, through the basal ganglia, to the thalamus, and back to the frontal cortex. This iterative-loop architecture might suggest that the distinctive feature of voluntary, or self-generated, actions is their dependence on our previous action history, as opposed to the immediate current external stimulation.

Follow-Up Questions

Jake Gavenas

If voluntary actions are indeed generated by a cortico-subcortical loop involving frontal areas, basal ganglia, and thalamus, how—if at all— should we revise our thinking about the neural precursors and causes of actions? To expand further, thinking about causation for cyclic connections is tricky. So activity in the entire loop, rather than in one particular area, might be considered the cause of voluntary action.

Amber Hopkins

It seems like we need experience to intend or act in the first place. We need other states and processes (sensory, perceptual, emotional, and so on) not only for motivation but also to have content to form intentions about and to then act upon. Are actions always in response to something? If so, could the "first cause" of our intentions and actions be experience? Many of these states and processes (sensory, perceptual, emotional, and so on) are out of "our" control or not "up to us." We do not have control over much of the external world, over at least some of our past experiences, over the behavior of others, and perhaps over some of our emotional states. So if our intentions and actions result from states and experiences that are out of our control, there may be less room for a "will" or "self" as a "first cause." Rather, experience would be the "first cause" to which we respond.

Tillmann Vierkant

What does it mean that the will has the "capacity to initiate endogenously those actions that correspond to our intentions"? Does that mean that if the agent has the intention to react to the stimuli presented, these reactions are endogenously initiated actions?

Eddy Nahmias

You start by saying, "We will consider that actions necessarily involve body movements, caused by muscle contractions." However, we also perform mental actions, such as recalling answers in a quiz or calculating a sum in our heads. Does what you say about neural processes apply to mental actions like these?

Replies to Follow-Up Questions

Patrick Haggard and Elisabeth Parés-Pujolràs

The reverse-engineering approach to voluntary action works by tracking backward along a causal chain that has voluntary action as its endpoint. This of course raises problems regarding first causes, which underlie the questions by Gavenas and Hopkins. Since we want to avoid positing uncaused causes, we need to be cautious regarding the starting point of this chain. While a reverse-engineering approach is useful for tracing the first steps back along the chain to primary motor cortex and to the medial frontal areas immediately anterior to it, it then becomes less useful for charting the next steps. This is because the metaphorical chain becomes a loop. Like many other areas of frontal and prefrontal cortex, the medial frontal cortex is a target of the subcortical loops that run through the basal ganglia. The function and operation of these is still debated, but they can be viewed as a form of flexible system for selecting actions from contexts. Multiple cortical areas input to the striatum.[7] From here, the balance of excitation and inhibition in pathways through the basal ganglia nuclei to the thalamus determines the strength of a drive back to the frontal/prefrontal cortex, producing the next action. Research on the neural correlates of intention and the neural precursors of action has tended to focus on some specific areas of this network rather than on the complex interplay between them. A focus on the dynamic interaction between areas in this distributed network might prove a valuable avenue for future research in voluntary action.

This loop-like arrangement determines the flow of voluntary actions in a number of ways. First, because the basal ganglia loop integrates information from such a wide swath of cortex, each action will depend on a wide range of features of the current context, and not just on a single imperative stimulus. Second, because the basal ganglia loop allows one action to flow from the one before, it provides a powerful focus for reinforcement learning—an idea surprisingly absent from

---

[7] One of the nuclei of the basal ganglia. The basal ganglia are featured in Figures 1b and 3 in the Brain Maps section.

discussions of voluntariness. Hopkins asks whether intentional action[8] begins with an experience. Our answer would be that all intentional actions involve some antecedent context, which will include both perceptual and motor experiences. What, then, does intention add to that antecedent context? If our voluntary actions follow *directly* from antecedent contexts, are they truly voluntary?[9] One answer to this question simply insists on the range and complexity of contextual factors influencing human action. This presumably relates to the wide range of information that the basal ganglia loops receive from the cortex: most cortical areas project to the striatum. However, invoking the complexity of contextual representations may leave less need for a theoretical construct of intention. A second possible answer might point to the strong limbic and motivational inputs to the striatum.

The striatal microcircuitry allows motivational areas such as the orbitofrontal cortex to influence the projection of contextual information through the basal ganglia pathways. Models of basal ganglia function that emphasize selection between alternative actions can thus point to motivational sources of choice. This architecture might account for contextual and state-dependent action preferences and could help in answering Vierkant's question. Vierkant raises the scenario of forming an intention to respond to an external stimulus. Once the stimulus occurs and the action is triggered, should one think of the action as endogenous or exogenous, as intentional or as reactive? We argue that actions that are produced in response to a stimulus can be intentional if they follow from a previous decision to respond to *that* stimulus *that* way. For example, when a runner is waiting for a gunshot to start the race, his reaction crucially depends on the fact that he previously decided to respond to that specific stimulus in a very specific manner. Importantly, he could have decided to do otherwise. This is sometimes referred to as a *prepared response*. Prepared responses differ crucially from reflex actions. Reflex actions are reactive in the strong sense that a particular stimulus invariably results in some bodily reaction regardless of a person's willingness for that movement to occur. In

---

[8] See Chapter 1 by Yaffe on intention.
[9] See Chapter 3 by Hieronymi on voluntary action.

contrast, a prepared response involves a decision regarding the particular circumstances under which the action will or will not occur.

Finally, we would like to address the first question from Nahmias about mental actions. The reverse-engineering approach has the virtue of beginning with the objective facts of a muscular contraction. Reasoning back from effects to causes seems less risky if we can establish clear facts about the effect. This approach struggles with purely mental actions, which may not have any overt, measurable counterpart at all. We know about purely mental actions only from our own subjective experience. Recent neuroscientific claims of mind-reading purport to have external, objective access to mental actions—but this can be at best a proxy, which requires additional grounding in subjective reports. So while we may believe that the processes generating mental actions may be broadly similar to the processes generating physical actions, the evidence supporting the two assertions would be quite different.

17

Does the will correspond to any clearly delineated brain area or activity?

Gabriel Kreiman

We know almost nothing about where, when, and how the will[1] is encoded in the brain. However, all the evidence to date suggests that the will should correspond to activity in the brain, not in the toes, nor are we "following our heart," nor is the will clearly correlated with activity in any other part of the body different from the brain. Clearly, not all the brain is involved in the will. For example, people who suffer a lesion in primary visual cortex have a scotoma, a circumscribed region of the visual field where they are blind, but do not manifest any problem whatsoever with their will. By a similar process of elimination based on lesion studies, it is possible to rule out the involvement of many brain areas in the will. Therefore, despite our general ignorance about the neuronal mechanisms involved in the will, the answer to this question should be "probably yes," the "probably" depending on the exact definition of "clearly delineated" in the formulation of the question.

In contrast to these lesion studies that leave the will intact, there are no clear studies showing that lesions (or other conditions) could lead to a complete abolishment of volitional decisions. It is not even entirely clear how exactly one would detect such a condition. A possible test might be to ensure that a subject can adequately respond to the instruction "Raise your right hand whenever I say GO" but fails completely to respond under instructions that leave the *when* and *what* aspects of

---

[1] See Chapter 2 by Hieronymi on the will.

Gabriel Kreiman, *Does the will correspond to any clearly delineated brain area or activity?* In: *Free Will.* Edited by: Uri Maoz and Walter Sinnott-Armstrong, Oxford University Press. © Oxford University Press 2022. DOI: 10.1093/oso/9780197572153.003.0017

the action up to them,[2] as in "Raise your left or right hand whenever you want within the next 30 seconds." Perhaps the closest examples are cases of reduced volition: severe substance addictions, obsessive-compulsive disorders, or severe depression. Addictions and obsessive-compulsive disorders are cases where subjects are largely unable to stop certain behaviors. (A person may "want" to stop washing their hands every five minutes, but they can't help themselves.) In contrast, severe depression connotes a large inertia toward volitionally initiating actions. The impairment in stopping certain actions (addiction) or initiating actions (depression) seems to point to disorders of volition. A few other terms, such as "apathy," "abulia," and "akinetic mutism," are also sometimes used to connote different degrees of inertia, indecisiveness, or general lack of initiative toward actions or speech. The exact neural structures and mechanisms involved in these conditions remain poorly understood. However, the preliminary evidence that we have makes it clear that these conditions are rooted in specific sub-circuits in the brain. Therefore, although still ill-defined, these conditions also suggest that the will corresponds to specific brain activity.

The closest the field has gotten to delineating specific brain activity associated with the will is to scrutinize the activity of neurons in tasks where an animal (human or not) performs volitional decisions. Several neurophysiological studies have shown that neuronal activity in specific brain regions correlates with, and often precedes, the time at which a volitional decision is executed. Studies in macaque monkeys have shown neuronal correlates of the will in supplementary motor areas (Romo & Schultz, 1992), putamen,[3] and parietal areas (Maimon & Assad, 2006). In some cases, the neuronal responses also precede the subjective report of when the decision was made, and neuronal correlates of the will preceding awareness have been described in humans in the anterior cingulate, supplementary motor area, and pre-supplementary motor area[4] (Fried, Mukamel, & Kreiman, 2011).

[2] See Chapter 18 by Triggiani & Hallett on the neural processes of deciding what, when, and how to move.

[3] The putamen is part of the dorsal striatum, the striatum being one of the nuclei of the basal ganglia. The basal ganglia are featured in Figures 1b and 3 in the Brain Maps section.

[4] The pre-supplementary motor area is the anterior part of supplementary motor area (SMA). The SMA is featured in Figure 2b in the Brain Maps section.

Interestingly, there is also convergent evidence showing that electrical stimulation in supplementary motor areas and in parietal areas can elicit the initiation of volitional decisions (Desmurget et al., 2009; Fried et al., 1991). In other words, upon invasively injecting electrical currents in these regions, subjects reported an urge to move, or the volitional decision to execute movement. These experiments are distinct from stimulation studies that elicit reflex or non-volitional actions. For example, a doctor may trigger a leg movement by tapping the right location in the knee, but in these cases, subjects report no sense of agency or volition.

Some of the studies cited earlier (e.g., Maimon & Assad, 2006) directly compared neuronal responses during the same decision task under reactive conditions (the subject needs to respond to a cue) versus volitional conditions (the subject initiates the same movement but without any cue). The findings suggest that the same neurons are involved in both cases, but the specific timing and pattern of activity differ. These results, exciting as they are, should not be interpreted as a complete delineation of the neuronal correlates of willed decisions. In particular, most studies focus on one single type of volitional decision. It remains unclear whether the same neural circuits would invariantly represent volitional decisions irrespective of the action specifics (e.g., do choices like raising the right or left hand, deciding on an ice cream flavor, or selecting a particular shirt all share the same neuronal mechanisms?). Much more work is needed to elucidate these neural circuits, and there may well be multiple overlapping circuits involved in volitional decision-making.

According to these preliminary findings, the will is but one specific manifestation of decision-making. We speculate that there is a continuum of decisions with different degrees of freedom,[5] from externally raising a subject's arm while they are sleeping (no will) to telling them to immediately raise their arm at gunpoint (minimal will) to a task where the subject is asked to raise an arm as soon as possible in response to an auditory tone (the subject has complete freedom but will usually comply with the instructions) to a task where subjects are

[5] See Chapter 4 by Sinnott-Armstrong on freedom and Chapter 7 by Hall & Vierkant on free will in degrees.

instructed to raise whichever arm they want whenever they want (most will). We conjecture that in all of these decision conditions, except the no-will case, the same neuronal decision-making mechanisms will be involved. Those neuronal circuits will be modulated and influenced by different brain processes (by fear cues in one case, by sensory cues in another, etc.). According to this conjecture, regardless of these task-specific modulations, the will is expressed through a common set of clearly delineated, though still largely unknown, neuronal circuitry.

Follow-Up Questions

Walter Sinnott-Armstrong

You say that "neuronal correlates of the will" have been found in the putamen and parietal areas in macaques, in the anterior cingulate in humans, and in the supplementary motor areas in both species. But what exactly are these correlates of? To what extent are they correlates of what Hieronymi identifies as the will in her chapter in this volume?[6] If they are correlated to a considerable extent, of which of the two pictures of the will that Hieronymi distinguishes are they correlates? If they are correlates of neither of these pictures of the will, does that show that philosophers and neuroscientists are talking about different things when they discuss "the will"?

Adina Roskies

Your explanation focuses on the overlap between the decision-making circuits in volitional and reactive actions, or where the characteristic activity is found. But is there a difference in the internal dynamics or how these circuits are activated that can distinguish willed from unwilled motor actions?

---

[6] See Chapter 2 by Hieronymi on the will.

Silvia Seghezzi and Patrick Haggard

You propose different example cases of reduced volition, like addiction, obsessive-compulsive disorder, depression, and inertia. Is it possible to find a common dysfunctional ingredient among these cases? If so, could a neural substrate, common to different cases of reduced volition, be informative in understanding the neurofunctional mechanisms of volition?

Replies to Follow-Up Questions

Gabriel Kreiman

Too many interesting and smart questions, too few words to answer them. Sinnott-Armstrong highlights the central issue of whether our usage of the word "will" coincides with that of others. Unfortunately, there are probably as many definitions of "will" as the number of philosophers and neuroscientists interested in the matter (e.g., Professor Hieronymi starts her chapter by admitting that the definition is controversial).[7] To make matters even more complicated, most of those are *verbal definitions* that cannot be readily compared to each other and do not lend themselves to concrete empirical evaluation. To make progress, the field needs an empirically *falsifiable* hypothesis (an idea that can be proven wrong by an experiment) and *operational* definitions, written in terms of measurable variables like reaction times, button presses, action potentials, etc. We hope that this new, exciting journey and invitation for a dialogue between neuroscientists and philosophers will help us jointly come up with better empirically testable definitions and hypotheses.[8]

We *operationally* use the word "will" in the following sense. Consider an experiment wherein there are two identically valued options (e.g.,

---

[7] See Chapter 2 by Hieronymi on the will.
[8] For additional discussions of this matter, see Chapter 9 by Roskies on neuroscientific evidence about free will and Chapter 26 by Hopkins et al. on consensus on consciousness and volition.

raising your right hand or your left hand), and the subject (human or not) chooses one or the other, in a *seemingly* random fashion, upon repeating the experiment multiple times. This operational definition matches our intuitions and the maximum common denominator of philosophical definitions. A rigorous definition will come only after we understand the neuronal circuit mechanisms underlying volitional decisions.

A tangential but essential point in which we reject some of the ideas put forward by Professor Hieronymi is the notion that the will is the exclusive possession of humans. Unfortunately, such anthropomorphic notions are pervasive in the form of statements like "Humans are the only creatures that have X," where X can be language, wills, or Nike shoes. There is no scientific evidence of discontinuities in evolution. Until proven otherwise by rigorous scientific data, we should assume that non-human animals also have a will (and also language-like mechanisms of communication, but we can safely assume that they do not have Nike shoes). Whether the *C. elegans* worm has wills or not and when wills arose in evolution is a fascinating question that we cannot answer right now. However, our working hypothesis is that non-human animals have wills too.

Francis Crick (1994) argued that we will not solve the problem of the neuronal mechanisms of consciousness if we get entangled with purely verbal definitions. In a similar vein, let us assume that we share some approximate intuition of what free will is and continue to investigate the relevant neuronal mechanisms. Another vital question, posed by Roskies, pertains to the mechanisms of *willed* versus *unwilled* actions. For the sake of addressing this question, let us remove from the discussion certain *reflexes*, such as the knee-jerk reaction when a doctor taps the knee (which is orchestrated by specialized spinal cord mechanisms) or moving the hand away from a painful stimulus like fire. Let us also exclude special cases like breathing, which can be volitionally controlled, but only to a limited extent. The simplest hypothesis that we should entertain until proven wrong is that *everything else is a willed action*. Hence, we propose that the essential mechanisms underlying *all decisions share common neural mechanisms*. Consistent with this hypothesis, neurophysiological studies comparing cued versus volitional movements show that the same neurons are active in both cases (e.g., those in parietal area 5; Maimon & Assad, 2006). This shared mechanism should not be taken to imply that different decisions are identical. For example, consider choosing what to eat at a restaurant, choosing what to say during a conversation, or choosing to raise the right or left hand. Different choices integrate different inputs (smells, history, auditory inputs, visual inputs), and many choices produce different motor outputs (sounds, eye movements, arm movements). There are neural circuits that convey visual inputs and others that convey auditory inputs. There are circuits responsible for moving the arms and other circuits for producing sounds. However, these distinctions are arguably less interesting for understanding the core neural mechanisms of the will and are not in contradiction with the notion of a common core decision mechanism.

Now consider two possible experiments:

Experiment 1: Tap a key with either the right hand or left hand (Fried et al., 2011).

Experiment 2: At gunpoint, you can tap a key with either hand, but if you use your right hand, the investigator will kill you. (Disclaimer: this is not a real experiment; it is not approved by an Institutional Review Board; do not try it at home.)

Continuing with the most straightforward hypothesis outlined earlier, we speculate that both experiments involve *willed* actions, and both experiments share the same neural circuits for making decisions. Those circuits receive inputs from many other brain areas. One of those other inputs provides a powerful signal in Experiment 2 but not in Experiment 1. That strong survival signal modulates the core decision mechanism to ensure that the left-hand decision wins. One can imagine a continuum of such experiments with different values for the right-hand and left-hand choices (due to differential rewards or because the subject is left-handed or because the right hand is tired or hurt or because history effects lead to novelty-seeking after 20 trials of using the right hand). Different inputs would modulate the core decision machinery in all of these cases. Nevertheless, we *hypothesize* that they would all be willed actions engaging the same underlying mechanisms. We would be happy to be proven wrong by empirical neuronal findings.

Taking this idea one step further, it is tempting to speculate that the different clinical conditions involving "reduced volition," such as obsessive-compulsive disorder, depression, or addictions, share a common dysfunctional ingredient (here, copying verbatim the words used by the lucid question by Seghezzi and Haggard). This hypothesis opens up an exciting line of inquiry to evaluate what that common ingredient is. Any progress along these lines could help us better understand the neuronal mechanisms of free will and could have important clinical implications as well.

18

How are the neural processes for deciding *when* to move similar to and different from those for deciding *what* or *how* to move?

Antonio Ivano Triggiani and Mark Hallett

When awake, persons are making movements all the time. For any one movement, it is possible to ask how and why the brain made that specific movement at that specific time with that specific synergy of muscles: *what*, *when*, and *how*. Of course, these features are intertwined and evolve from influences originating in internal drives and environmental stimuli. Their complexity and interrelationships make the identification of the neural processes underlying *when*, *what*, and *how* to move difficult.

Voluntary movements can be distinguished as spontaneous (self-initiated at the time of the movement) and triggered (for example, an immediate response to an external visual stimulus), and the *when* question is relevant only for the spontaneous ones.[1] If we want to isolate the neural process underpinning the decision "*when* to move," it is important that the experiment is based on a simple and well-known act. In this way, the subject can freely decide when to initiate the movement without being distracted by instructions or the complexity of a decision about what to do or how to move. Libet-like experiments[2] represent

---

[1] See Chapter 3 by Hieronymi on voluntary action.
[2] For further details about these experiments, see Chapter 19 by Bold et al. on arbitrary and deliberate decisions and Chapter 28 by Lee et al. on the timing of mental events.

Antonio Ivano Triggiani and Mark Hallett, *How are the neural processes for deciding* when *to move similar to and different from those for deciding* what *or* how *to move?* In: *Free Will*. Edited by: Uri Maoz and Walter Sinnott-Armstrong, Oxford University Press. © Oxford University Press 2022. DOI: 10.1093/oso/9780197572153.003.0018

classic examples of this kind of task, and the Bereitschaftspotential (BP) is recordable in this condition with EEG electrodes over different motor areas. In particular, the early BP (or BP1) seems generated by the pre-SMA (pre-supplementary motor cortex), proper SMA,[3] and lateral premotor cortex bilaterally, while the late BP (or NS' or BP2) is generated by the primary motor cortex (M1) and lateral premotor cortex (predominant over the contralateral cortex) (Shibasaki & Hallett, 2006). According to the 10-10 International System for the placement of electrodes, the BP is clearly recordable using the electrodes Cz (center), C3 (left), and C4 (right). Functional magnetic resonance imaging (fMRI) studies of similar simple movements also show these areas, as well as significant activation in the ipsilateral cerebellum and weaker activations in a variety of areas.

In a variation of the Libet experiment using fMRI, spontaneous movements were contrasted with visually triggered movements. Initially spontaneous movements were made, and each movement triggered a light to go on. The sequence of lights was recorded, and then in a second run, the lights were played back to subjects to trigger the movements. Hence, the movements and the precise timing were the same, but only in the first run was there a decision of when to make the movements (Deiber, Honda, Ibañez, Sadato, & Hallett, 1999). Results showed increased activity in the pre-SMA and cingulate cortex for the self-initiated condition with respect to the externally triggered condition. Despite some divergences in results, the network for the *when* circuit seems to be the following: the final inputs for motor execution are received by the M1 from the pre-SMA, which in turn elaborates information from the prefrontal cortex and the basal ganglia.

The ideal *what* component could be isolated by instructing the subject to freely choose among different possibilities, but with cued timing, avoiding the *when* component. One experiment of this type compared free choice to instructed choice. Older studies suggested that the dorsal prefrontal cortex (and not the pre-SMA) was important for this decision, but more recent fMRI studies during free selection of *what* seem to point at the pre-SMA and the cingulate cortex.

---

[3] The pre-supplementary motor area is the anterior part of the supplementary motor area (SMA). The SMA is featured in Figure 2b in the Brain Maps section.

Specifically, free selection of right or left index finger flexion involved both these areas, while specification of a button (left or right) in a cued task showed activation only in the pre-SMA (Hoffstaedter, Grefkes, Zilles, & Eickhoff, 2013). While there are some discrepancies, the *what* circuit shows similarities with the *when* circuit. The nature of the *what* and *when* decisions in these simple circumstances may indeed be similar but certainly differ with decisions of greater complexity. Choosing what to do when the outcomes have different consequences would bring other areas into action. For example, when there are different levels of reward, dopaminergic influences operating through the basal ganglia will influence the choice. Making a decision of *what* when the outcomes are equal is called "picking" (rather than "choosing"),[4] and this type of decision might be more difficult for the brain or may just be stochastic.[5] It is likely that almost any part of the brain might be involved in the selection of what to do for different movements.

The *how-to-move* condition is associated with the concept of "praxis"—i.e., the ability to perform voluntary, skilled movements. Each skill, generally a complex movement, requires a separate motor program that is acquired by practice. Historically, the first information about the relevant brain circuitry for praxis came from patients with brain lesions who lost the ability for skilled movements. This deficit is called apraxia, and the most common type is ideomotor apraxia with loss of the spatial-temporal properties of a motor program. For example, someone might just move their hand up and down when asked to show how to use a key in a lock. Lesions that lead to apraxia are in the left parietal and premotor regions or their interconnection for both right or left hand movements in right-handers. Studies of praxis using fMRI and EEG confirm the relevance of these regions. Experiments typically distinguish between transitive gestures (that involve the use of a tool) and intransitive gestures (without tools but with semantic connotation, e.g., pointing at something). The execution of transitive gestures involves a stronger activation of the posterior parietal cortex, specifically the inferior part, the intraparietal sulcus, and anterior

---

[4] See Chapter 19 by Bold et al. on arbitrary and deliberate decisions.
[5] See Chapter 29 by Liljenström on stochastic neural processes.

intraparietal cortex (Fridman et al., 2006). The *how-to-move* network therefore differs from the network of the *when* and *what*.

For most movements, presumably the decisions would logically have to be in the order of *what* to do, *how* to do, and then *when* to do. In a common circumstance, I might decide that I need to brush my teeth, then I will need to plan to use the toothbrush in the appropriate manner, and finally decide when to do it. Any or all parts of the brain would be relevant at any time selecting what to do. Then the dominant parietal-premotor system helps with the planning of the motor control pattern, which is activated at a specific time with the premotor and supplementary motor influences on the motor cortex.

Follow-Up Questions

Walter Sinnott-Armstrong

Suppose a robber offers two options, "Your money or your life." The victim does not freely choose what to do, since she is coerced into handing over her money. Nonetheless, she still seems free within limits to choose how to hand over her money (quickly or slowly, with her right hand or her left hand) and also when to hand over her money (immediately or a minute later, after pausing to think about her options, but not next week). More generally, can a person be free to choose what to do without being free to choose how to do it or when to do it? Or be free to choose when to act without being free to choose what act to do or how to do it? Or be free to choose how to do something without being free to choose what to do or when to do it? If so, can your distinctions among *what*, *how*, and *when* to act help us avoid confusions in discussions of free will?

Adina Roskies and Jye lyn Bold

In the last paragraph you discuss the logical order of decisions (*what* to do, *how* to do, and *when* to do it) as if these decisions had to be made serially. But must they be, and can we find signatures of that timing in the brain? How do these decisions relate to the psychological term "intention"? And why is it logical for the order of decisions to be *what* to do, *how* to do it, and *when* to do it? In particular, for more complex decisions, would the order not vary greatly, potentially with parts repeating?

Tillmann Vierkant

The distinction between spontaneous and triggered voluntary movements is about different triggers for the movement, but the *what* choices seem to be about forming intentions rather than triggers for movement. Or is a choice based on evidence triggered, and only picking is spontaneous?

Replies to Follow-Up Questions

Antonio Ivano Triggiani and Mark Hallett

A robbery in which a victim is threatened with the classic decision, "Your money or your life," will be the scenario of our answer. Furthermore, let's say that the victim had been mugged in the past to mimic the background information often given to subjects in experimental studies of this type.

1. The distinction among different factors in the investigation of free will

We can separate the three items, *what*, *how*, and *when*, in terms of thought and underlying neural mechanisms. Thus, it is reasonable to conclude that free will relative to each factor is independent. The victim must choose what to do: giving the money, fighting, or trying to flee. The decisions of how to handle the money and when to do it, if she were to decide on that choice, are separate. Thus, every experiment aimed at investigating free will can address each of these three items and avoid some confusion. Regarding the investigation of free will in this specific situation, it is worth noting that if the victim had a prior similar experience, this experience and outcome might influence her current actions.

Interestingly, the neural processing of the actions performed under coercion might be underpinned by reduced brain activity. An experiment showed that acting when obeying orders activates a brain process that is closer to a passive movement than to a completely voluntary action (Caspar, Christensen, Cleeremans, & Haggard, 2016). Thus, when an agent does not act completely freely, this can be identified in correlative brain activity. Our interpretation of the situation is consistent with the "compatibilist" point of view.[6]

2. The sequence and the timing of specific items of action and their relationship with the intention

The order of a sequence we proposed comes from thinking about the items in a logical order, and in many situations, it certainly might be in the proposed way. However, since the factors of *what*, *how*, and *when* are independent, they could be in any order. For example, the victim could decide how to give the money before actually choosing that alternative. Also, a distal intention of *when* could also be made in any order, although the proximal intention would ordinarily be last. (However, even this is not necessarily true, as will be discussed.) Additionally, it is reasonable to think that some decisions might be repeated and changed prior to the action.

There is an experimental paradigm, the "timed response paradigm" (see, as an example, Favilla, Gordon, Hening, & Ghez, 1990), that manipulates the timing of the *what* decision in relation to the timing of the "proximal *when*" decision.[7] Movements are trained to occur on the fourth of a series of rhythmic stimuli, and, after practice, subjects are highly accurate in timing. Information about what movement to make can be given at various intervals prior to the movement itself. In such an experiment, it can be determined how long it takes for the *what* information to be fully integrated so that the movement is correct. The results show that movements can be made even in the absence of an

---

[6] For more on compatibilism, see Chapter 4 by Sinnott-Armstrong on freedom and Chapter 5 by O'Connor on free will.
[7] See Chapter 28 by Lee et al. on timing mental events.

apparent *what* decision (or only a partially completed *what* decision). Without knowing *what* to do, subjects created a response with characteristics that reflected their expectations. Moreover, those expectations seem to be influenced by past experiences.

3. The influence of a trigger on a choice

The trigger for a movement can be internal (self-initiated) or external, both of which involve the *when* issue. When internal, it is part of the free-will decision for that movement and can be considered spontaneous. When external, it is much less free, although in some circumstances the person can decide (freely) whether or not to respond and even possibly exactly when to respond. In an experimental situation, the subject may make a distal decision to always respond to an external trigger, so then there is really no free choice with each stimulus. The *what* decision generally relates to a more distal intention and is not a trigger for the movement. The *what* decision can be instructed, picking, or choosing. True picking is not that common because in most situations the possibilities are unbalanced. For example, in the robbery, for most subjects, their life would be weighted more than some money. Additionally, as noted earlier, a coerced decision of handing over the money with the dominant hand requires less brain activation (François-Brosseau et al., 2009), and, likely, less energy expenditure, also making it an easier choice. Even in apparent picking situations, there can be unconscious biases that influence the decision. A study showed that, during a free-choice task, the response was faster and more frequent when the response was compatible with an unseen prime compared to when the response was incompatible (Furstenberg, Breska, Sompolinsky, & Deouell, 2015). This shows an effect of unconscious bias. A person who survived an earlier mugging might find it easier to just hand over the money the second time, believing that if she did so, she might well actually survive again. Past experiences change the probabilities of responses in specific situations, but the responses can still be considered free.

19

How are arbitrary and deliberate decisions similar and different?

Jye lyn Bold, Liad Mudrik, and Uri Maoz

Arbitrary and deliberate decisions are similar in some ways. The most obvious might be that the decision-making processes in both result in a commitment to a plan of action. However, the two types of decisions might also be different in several important ways: their underlying cognitive processes, their associated behaviors, and their underlying neural mechanisms. We will here focus on the last of these potential differences. In particular, we will ask whether any differences in the underlying neural mechanisms pertain to the generalizability of the Libet results beyond arbitrary decisions.

As a first step, let us define what we mean by arbitrary and deliberate decisions. The key distinction here, originally suggested by Ullmann-Margalit and Morgenbesser (1977) with respect to what they termed *picking* vs. *choosing*, lies in the role of reasons in these decisions. Arbitrary decisions are unreasoned; they cannot be resolved by deliberation. A classic example is reaching for a carton of milk at the supermarket, among a shelf full of seemingly identical cartons (same brand, fat content, expiration date, etc.). Having no reason to prefer one over the other, you might randomly pick one (or take the one closest to you). This would be an arbitrary (or picking) decision. However, if the milk with the fat content you prefer has an earlier expiration date than that of another fat content, for example, you might deliberate, say, dietary preferences vs. shelf life. That reasoning process would render this a deliberate (or choosing) decision.

Interestingly, the neuroscience of free will has historically focused on arbitrary decisions. The seminal work of Benjamin Libet (1985) targeted such decisions with respect to the Event Related

Jye lyn Bold, Liad Mudrik, and Uri Maoz, *How are arbitrary and deliberate decisions similar and different?* In: *Free Will.* Edited by: Uri Maoz and Walter Sinnott-Armstrong, Oxford University Press. © Oxford University Press 2022. DOI: 10.1093/oso/9780197572153.003.0019

Potential (ERP) component called the readiness potential (RP). This component—originally reported as the Bereitschaftspotential by Kornhuber and Deecke (1965)—is an increased EEG negativity recorded above the pre-supplementary motor area[1] (a brain region associated with motor planning), leading up to movement onset. In the early and mid-1980s Libet was intrigued by the possible relations between the RP and subjects' conscious experience of deciding to move. He accordingly instructed subjects to flex their right wrist or fingers whenever they felt the urge to do so. At the same time, they sat in front of a screen with a rotating dot and were asked to report its position when they first felt the urge to move. Strikingly, subjects' RPs started about 500 ms before movement onset, and—most important—roughly 300 ms before their reported urge to move. The Libet (1985) experiment, famously taken as evidence against free will, has been replicated and expanded many times, typically using arbitrary rather than deliberate decisions. This is probably because arbitrary decisions are viewed as the most "distilled" type of volitions and endogenous actions—devoid of any kind of external influences by reasons and values—and thus perhaps less confounded and even more free. On the other hand, typical real-world decisions are often deliberate. And such deliberate decisions are also arguably more relevant to the free-will debate. For example, free will is often viewed as the capacity that enables moral responsibility. But responsibility tends to be associated with deliberate decisions—purposeful, reasoned, and bearing consequences.[2]

Therefore, it is critical to know whether the Libet results can be generalized to deliberate decisions. Is the RP a general index of internal decision-making, or is it specific to arbitrary decisions? And, more generally, are the neural correlates of arbitrary and deliberate decisions similar or different in a manner that matters to free will and to moral responsibility?

Recently, several studies compared the neural activations associated with these two types of decisions, sometimes yielding conflicting results. The first, Maoz, Yaffe, Koch, and Mudrik (2019), used

---

[1] The pre-supplementary motor area is the anterior part of supplementary motor area (SMA, or proper SMA). The SMA is featured in Figure 2b in the Brain Maps section.
[2] See Chapter 14 by Yaffe on responsibility.

a donation paradigm to probe both arbitrary and deliberate decisions. Participants were presented with two non-profit organizations. In one experimental condition, they were asked to press either the right or left button to indicate their preference to donate $1,000 to the left or right organization on the screen (and such a donation was indeed made). In the other condition, they again pressed right/left, but their decision did not affect the outcome (both organizations received $500 regardless of which button was pressed). The RP was found in the latter case but was strikingly absent in the former. The lack of RP in deliberate decisions therefore suggests that the RP might not be a general index of decision-making; it also suggests that arbitrary and deliberate decisions may differ with respect to the relations between neural activity and one's conscious experience of deciding.

Two other studies potentially support differential neural correlates between arbitrary and deliberate decisions. In the first, Travers, Friedemann and Haggard (2020) had participants take part in a reinforcement learning paradigm, where they learned, through trial and error, the optimal time to act. The researchers found that the amplitude of the RP was smaller for the initial trials, when the participants were learning the task, than for later trials, when they were already trained. The researchers discuss their task in relation to arbitrary versus deliberate decisions. However, this distinction was not directly manipulated in their task. And, further, it is debatable whether more deliberation took place earlier in the task—while the participants were trying to learn the underlying regularity of the task in a reasoned manner—or later in the task—when movements where more planned and deliberate. Notwithstanding, to the extent that early versus late trials relate to the distinction between arbitrary and deliberate decisions in this task, the authors found different RP amplitudes between the two decision types.

The second study, Khalighinejad et al. (2019), did not investigate the RP. Instead, it focused on another EEG component—across-trial variability—which decreases more before self-initiated than before externally triggered actions. The task was composed of self-initiated blocks, where participants decided for themselves on each trial whether to wait a potentially long time for a large reward (whose delivery time followed an exponential distribution with a maximum of 1 minute) or skip the wait and settle for a small reward. In the externally triggered blocks, the participants were instructed when and whether to skip the wait. Interestingly, EEG across-trial variability decreased more markedly before self-initiated compared with externally triggered "skip" actions, especially if the skip responses were rationed, encouraging deliberate planning.

In contrast, two other studies did not find differences between the RP in arbitrary and deliberate decisions. Verbaarschot, Farquhar, and Haselager (2019) presented a video game, *Free Wally*, where subjects had to help a caged whale on the top of a hill escape into the water. The game included friend and foe characters who would climb toward Wally from the left or right to add or remove bars on its cage. Participants were instructed to decide if Wally should spray water to the left or right side of the hill and wash away the characters on that side. Hence, on each trial, participants had to make a deliberate decision of *whether* to act, and if so, *when* to act and *what* to do (which side to spray). In the other condition, run on a different subject pool, a video game of shapes moving toward a star on a hill was used, where subjects would arbitrarily press right/left. It lacked any story line, strategy, or suspense. In other words, the what/when/whether decisions in this condition were more arbitrary. Despite the difference between the conditions, the RPs in the two conditions were fairly similar.

In another study, Parés-Pujolràs et al. (2021) instructed participants to monitor a stream of discrete visual stimuli (or evidence items) over time and decide whether or not to act. The evidence in each trial was either strong or ambiguous, reflecting more deliberate or more arbitrary decisions, respectively. However, they found no clear differences between the RPs in the two types of trials.

Finally, the results of another study, by Travers and Haggard (2019), fell somewhere in between these two positions. In this study, subjects decided whether to act or withhold action to accept or reject easy/difficult gambles that could translate into a small amount of money at the end of the experiment (about £2). In some trials, the expected win/loss values of the gamble were clear. In others they were hidden, and subjects had to arbitrarily guess. The authors did not find the typical RP for deliberate decisions (easy/difficult gambles) in the original EEG signals. However, they applied a PCA-like procedure (a statistical method meant to decorrelate observations) and manually selected the component that visually resembled an RP for further analysis. After this procedure, they did find an RP-like motor component for both easy and hard deliberate decisions, albeit both were smaller than the one evoked by guesses.

In sum, at present the literature is far from a consensus regarding whether deliberate and arbitrary decisions rely on different neural mechanisms. Three studies suggest they may, two others suggest they might not, and another yielded some support for both points of view. How can this be reconciled? One idea is to examine differences between the experiments that might explain the conflicting results. The tasks, for example, may have varied in the complexity of the visual input, in cognitive load, and in the actual consequences of the decision—and thus potentially the level of deliberation—among other possible differences. However, there were no clear differences in visual complexity or cognitive load between the studies supportive of and opposing differential neural mechanisms. In other words, while Verbaarschot et al. may have relied on more complex visual inputs and higher cognitive load, those of Travers & Haggard, Pares Pujolras et al., Travers et al., and Maoz et al. were arguably not that different in those respects.

Conversely, the consequences of the decision might be more different between supportive and opposing studies. Maoz et al., resulting in an actual $1000 donation to charities, is clearly the most consequential. The stakes in Khalighinejad were lower, but still included up to about £10 of bonus pay for the participants, if they completed the task optimally. (The reward scheme in Travers et al. was not clearly spelled out, so it is hard to gauge how meaningful the decisions were for the participants.) In contrast, the consequences for the participants in Travers & Haggard appear to have been only around £2. And in Pares Pujolras et al., participants' actions apparently resulted in about £1.5 additional pay. For Verbaarschot et al. the consequences were saving the whale or not rather than monetary ones; so, those depended on how important it was to the participants to save the whale.

Hence, there might be a gradient whereby the more meaningful the consequences of the participants' decisions the smaller the RP, with meaningful enough consequences leading to the RP possibly disappearing altogether. Such results are in line with the computational model described in Maoz et al. (2019). Nevertheless, more research is needed to better substantiate this hypothesis and to clarify how different—if at all—arbitrary and deliberate decisions are with respect to the free-will debate.

Follow-Up Questions

Walter Sinnott-Armstrong

You define arbitrary vs. deliberate decisions in terms of the role of reasons in these decisions. What exactly do you mean by reasons here? Which kind of reason are you talking about? How is your notion of reasons related to the philosophical distinctions that I draw in my answer to the question "What are reasons?"[3] in this volume?

Jake Gavenas

Deliberate decisions have been extensively studied in such fields as neuroeconomics. To what extent are arbitrary decisions just a "special case" of deliberate decisions (i.e., when the values of two options are equal or very low), as opposed to a qualitatively different kind of decision altogether?

Amber Hopkins

If it became clear that the readiness potential is generally present before deliberate decisions, would that threaten a role for consciousness or conscious intentions in human decisions and actions? Alternatively, if it became clear that the readiness potential is not present for deliberate decisions, would that leave a place for consciousness in our decisions

---

[3] See Chapter 15 by Sinnott-Armstrong on reasons.

and actions? Thus, in other words, would the absence or presence of the readiness potential in deliberate decisions have a bearing on the role of consciousness in decision-making?

Tillmann Vierkant

Compare two cases. In the first, you have to choose between two pints of milk, when one of them is past its "best before" date and you would never consider buying produce that is past its "best before" date. In the second, you again have to choose between two pints of milk, and one of them is past its "best before" date, but this time you would never consider buying produce past its "use by" date. Both of these decisions are "deliberate" in your sense. However, in both cases one might worry whether the deliberate choice is either really deliberate or really a choice. The first decision requires only that you apply a simple rule that you were already committed to. Does this really require deliberation, and is your decision in this case really even a choice at all even though your rule clearly mandated one option and you might even have automatized following that rule? In the second scenario the choice requires working out whether "best before" means the same as "use by." This clearly does require deliberation, but once you have worked it out there are then two quite different scenarios, and one of them seems to collapse into scenario 1. If it turns out that the two terms do mean exactly the same, then at the moment of choice this scenario seems to be just the same as scenario 1 in that your choice is now simply mandated by the rule. Do you think that in this case the choices in the two scenarios are deliberate choices in the same sense, or does it matter that there had been deliberation about the terms before the choice in scenario 2?

Replies to Follow-Up Questions

Jye lyn Bold, Liad Mudrik, and Uri Maoz

Following the distinctions of Sinnott-Armstrong, deliberate decisions (or choosing) may be related to explanatory, practical, and epistemic

reasons. Arbitrary decisions (or picking), in contrast, appear to relate only to explanatory reasons, if at all. Consider the following example as an explanation. You are asked to raise one of your hands for
no reason or purpose, and you raise your right hand. When asked
why you did that, you might come up with an explanation (e.g., "I am
right-handed"; "I felt like it"). But you would be hard-pressed to answer why you believe raising your right hand was the correct act, and
even more so why you should raise your right hand as opposed to your
left in this situation. Compare that with a new scenario, where again
you are asked to raise one of your hands, but now raising your right-
hand means donating $1,000 to save the whales in the northern Pacific
Ocean and raising your left-hand means donating $1,000 to save the
Amazon Rainforests. This time, if you decide to raise your right hand,
you could explain why you did that (e.g., because your desire to save
the whales is greater than that to save the rainforests). You could also
explain why anyone should raise their right hand (e.g., because saving
animals is more important than saving trees) and why you believe this
was the correct act (e.g., because you think it is more important to save
whales than trees). In other words, deliberate decisions are determined
by differences in preferences among alternatives, and thus they rely on
the specific types of reasons mentioned here.

Generally, deliberate decisions get harder as the preferences
among the alternatives become more similar. In contrast, when the
preferences among the alternatives are symmetrical, or where one is
indifferent with regard to the alternatives, the decisions are termed
"arbitrary" (or picking; Ullmann-Margalit & Morgenbesser, 1977).
Thus, one type of arbitrary decision lies at the upper limit of hard deliberate decisions, as the preferences among the decision alternatives
become indistinguishable (or there being no overriding reason). This
could happen if one is deliberating between donating to two charities
which one supports equally. Such arbitrary decisions tend to be associated with especially long reaction times, which are consistent with a
deliberation process that is halted when it cannot converge. An additional type of arbitrary decision is when the decision alternatives are
identical—as in the example above of selecting among several identical
milk cartons in the supermarket. Here the reaction times tend to be
especially short, consistent with there being no deliberation process

among identical alternatives—no attempt to decide the matter based on reasons. A third type of arbitrary decision relies on indifference—for example, when one is powerless to realize one's preference (e.g., Maoz et al., 2019). Imagine you are faced with two buttons. You know that one is associated with a large reward and the other with a small reward, but you do not know which one is associated with which reward. You can make only an indifferent choice. As a whole, one could claim—together with Jake Gavenas—that arbitrary choices are specific (perhaps even limited) cases of deliberate choices when the alternatives are of equal value, are identical, or induce indifference. However, as discussed in this chapter, there is evidence that some types of arbitrary decisions might rely on different neural mechanisms than deliberate decisions, which lends some support to deliberate and arbitrary decisions being qualitatively different kinds of decisions.

In one of the studies discussed in the chapter, Maoz et al. (2019) found that a readiness potential (RP) was absent in deliberate but present in arbitrary decisions. This finding seems to support the Schurger model, which suggests that the RP stems from artificial accumulation of stochastic fluctuations rather than being an index of unconscious decisions (Schurger, Sitt, & Dehaene, 2012). Following this logic, the RP should not be found in deliberate decisions, as those are driven by values rather than stochastic fluctuations or other symmetry-breaking mechanisms. If, as Amber Hopkins inquired, we would find an RP for deliberate decisions, this would challenge this reasoning and support the idea that the RP might indeed index unconscious decisions, as it is present both in deliberate and in arbitrary choice. The above notwithstanding, it is important to note that the absence of an RP in deliberate decisions does not constitute positive evidence for the involvement of consciousness in deliberate decision-making; rather, it serves as negative evidence—i.e., it is evidence against the claim by Libet and others that there is no role for consciousness in decision-making and action preparation.[4]

Vierkant poses a question about a distinction between two other scenarios. In both cases he mentions there is a two-stage process.

---

[4] See Chapter 9 by Roskies on neuroscientific evidence about free will.

First you establish a rule ("Do not buy milk if it is past its best-before date" or "Do not buy milk if it is past its use-by date"). Then you test instances of milk cartons (or pints) to find out whether they conform to the rule. For the best-before rule, the determination is straightforward and mainly perceptual (ascertaining that today's date is earlier than the best-before date). On the other hand, for the use-by rule some calculations need to be carried out on top of the perceptual task: how long from the listed best-before date to a use-by date is reasonable for this milk carton? After that, similarly to the first scenario, the determination follows a mainly perceptual effort. One could debate whether the decisions that are termed perceptual are even decisions at all or are rather just perceptual judgments. However, establishing rules probably involves some deliberation. And so, although the immediate application of that rule to a specific decision might not require new deliberation, the decision as a whole, including establishing the rule, might nevertheless include deliberation. Moreover, in many cases one may also deliberate over the rules and consciously engage in the process of applying them. If you were to go to the supermarket and pick up a carton of milk whose date expired, you might think, "Why is it still on the shelf? I will surely not buy a product after its expiration date." Hence, retrieving the rule and consciously thinking about it might also be a form of deliberation. When the decision is especially easy, such deliberation may be minimal but still present by means of thinking about the rule and applying it. Therefore, if scenario 1 is a deliberate decision, we propose that so is scenario 2. Following Sinnott-Armstrong's chapter on reasons, we note that one might follow reasons in the two scenarios even if those reasons are not explicitly or consciously available.

20

How do higher-level brain areas exert control over lower-level brain areas?

Mark Hallett

This question has the implicit assumption that there are higher and lower levels in the central nervous system (CNS). The notion is somewhat vague and arises from the general idea that there is a hierarchy of command among brain structures. The cortex is the highest level; the spinal cord is the lowest level; and other parts are in between. Decisions are made at high levels and carried out at lower levels. A high-level command would be "Grasp that cup." This would lead to a lower-level command of the muscles in a particular synergy to carry out that command.[1] Most neuroscientists will think of it in this way, but there is an alternative view. All parts of the CNS area connected to each other and continuously communicating; hence, it is possible to conceive that all levels are equal since it is just one large network. For the purposes of this question, however, we will take the view that there are levels.

Muscles make movements, and they are under the direct and sole control of the alpha-motor neurons in the spinal cord (or brain stem for the cranial nerve muscles). Understanding the control of the alpha motor neurons is therefore crucial for the control of movement. They receive input from the spinal segment in which they are located and descending input from the brain stem and cortex. The spinal segment is complex, with many synaptic connections. Afferents come from skin, muscle, and joint and make synaptic connections leading to excitation or inhibition of the motor neurons. These influences are called "reflexes." The simplest example is the monosynaptic connection of the Ia afferent nerve ("one-A" afferent; a large-diameter sensory nerve)

---

[1] See Chapter 21 by Parés-Pujolràs & Haggard on intentional action.

Mark Hallett, *How do higher-level brain areas exert control over lower-level brain areas?* In: *Free Will.* Edited by: Uri Maoz and Walter Sinnott-Armstrong, Oxford University Press. © Oxford University Press 2022. DOI: 10.1093/oso/9780197572153.003.0020

from the muscle spindles directly on the motor neuron; this mono-synaptic connection is responsible for the tendon-jerk phenomenon. A more complex example is the flexion reflex, a polysynaptic reflex mediated by pain fibers that lead to withdrawal of the body part from a potentially damaging stimulus.

The principal descending pathways to the spinal cord come from the reticular formation, the reticulospinal pathways, and the corticospinal tract originating mostly in the primary motor cortex (M1 or Brodmann Area 4). Both the various reticular nuclei and M1 receive widespread inputs from all over the CNS, including from all the senses. Ordinarily the corticospinal tract gives most of the commanding descending signals, particularly for fine motor control. The reticular formation likely provides a stable foundation, including posture and balance. The corticospinal tract has a significant influence on the reticular formation. M1 is laid out in Brodmann Area 4 by body part; this is called the "homunculus." Each subregion of M1 controls a part of the body; it is not muscle by muscle—in fact, each neuron is connected to many muscles. Thus, the descending signal must be a complex mixture of excitation and inhibition to get the desired movement.

Since M1 and the corticospinal tract are so important, it is necessary to know how these are influenced. The influences on M1 and the corticospinal tract are thought of as both subcortical and cortical. The subcortical influences are the basal ganglia and the cerebellum. The main circuitry of the basal ganglia is cortex to striatum, striatum to pallidum,[2] pallidum to thalamus, and thalamus back to cortex. The circuit is importantly modulated by dopamine from the substantia nigra pars compacta.[3] Dopamine has multiple roles, including promoting movement and mediating the facilitatory effects of reward.[4] The main circuitry of the cerebellum is cortex to pons, pons to cerebellum, cerebellum to thalamus, thalamus to cortex. The cerebellum also receives direct sensory input from the spinal cord. Additionally, it is modulated

---

[2] The striatum and pallidum are part of the basal ganglia. The basal ganglia are featured in Figures 1b and 3 in the Brain Maps section.

[3] The substantia nigra pars compacta is part of the substantia nigra, which is part of the basal ganglia. The basal ganglia are featured in Figures 1b and 3 in the Brain Maps section.

[4] See Chapter 23 by Hopkins & Maoz on beliefs and desires.

by another network, the Guillain Mollaret triangle, from inferior olive to cerebellum to red nucleus[5] and back to inferior olive. The basal ganglia and cerebellum have some interconnections. The general function of the basal ganglia is to help initiate what to do and to help prevent what not to do and to provide a "motor energy" (mediated by dopamine) (Hanakawa, Goldfine, & Hallett, 2017). The general function of the cerebellum is to help control the precise timing of movements, including online control of accuracy (Bodranghien et al., 2016).

Cortical inputs come to M1 from all over the brain. Those from the back of the brain are conveying sensory information from visual, somatosensory, and auditory modalities. These signals appear to control responses related to events in the outside world, such as moving the arms to catch a ball. The signals from the front of the brain (prefrontal cortex) seem to relate more to influences internally generated in the brain, such as homeostasis and emotion. These signals would lead one to get a drink if one is thirsty or go after a piece of chocolate since that was pleasurable the last time. The back part of the brain, the parietal cortex, also seems to store the motor programs for skilled movements and works together with the premotor cortex to produce the signals that lead to these skilled movements (Wheaton & Hallett, 2007).

All these inputs are converging on M1. Its eventual output depends on the weight of the synaptic input at any one time—a very complex calculation. With the model of "levels," the inputs to M1 are the highest level; M1 (the main executive level) would be the next level down, then the brain stem, and then the spinal cord. The basal ganglia and cerebellum would be modulators at the executive level.

When and where does a volitional movement originate?[6] In a certain way of thinking, this may not be a sensible question. The brain is always active, and various signals are traveling around the networks. The motor cortex sends signals when it is sufficiently facilitated. For

---

[5] The Guillain Mollaret triangle (also known as the dentatorubro-olivary pathway) has as its corners the red nucleus (part of the midbrain), the inferior olivary nucleus (structure in the medulla oblongata), and the contralateral dentate nucleus (in the cerebellum). The midbrain, medulla oblongata, and cerebellum are featured in Figures 1b and 3 in the Brain Maps section.

[6] For further discussion on this question, see Chapter 16 by Haggard & Parés-Pujolràs on the neural processes that produce actions.

any particular movement, however, it might be possible to say from where in the brain the main influences arose. There is no one volition center. The sense of volition is a mental state, a quale, and exactly how it is generated and whether, when generated, it has any causal influence are important questions without clear answers (Hallett, 2016).[7]

Follow-Up Questions

Walter Sinnott-Armstrong

You discuss a hierarchy among brain areas, but another way of understanding higher and lower levels is in terms of generality or spatial scale. We might say that higher-level or larger-scale activity throughout the amygdala (or some other relatively large brain part) controls, influences, or causes our whole body to move away from the snake that we just saw. However, we do not usually ascribe such control or influence to any particular neuron inside the amygdala (or elsewhere). So which causal relations exist between an individual neuron (at a lower level of generality) and activity throughout a larger brain region (at a higher level of generality)? Can neuronal activity throughout a large brain area cause an individual neuron to fire? Can the firing of an individual neuron cause neural activity throughout an entire brain region? And which level has causal influence on whether our whole body moves away from the snake that we just saw? Or on whether we as a whole person choose to move away? In short, how do causal relations between these levels of generality work?

Jake Gavenas and Amber Hopkins

The CNS hierarchy might lend itself to the idea of a causal chain where "higher" brain regions influence "lower" brain regions more than vice versa (at least for motor output). However, in your

---

[7] See Chapter 12 by Block on consciousness causing physical action and Chapter 25 by Gavenas et al. on conscious control of action.

introduction you allude to another way of thinking about the brain, where "all levels are equal." How might we think about the neural causes of action under such a flat framework? And what might be the benefit of such an approach to neuroscience regardless of the question of causality?

Adina Roskies

Don't some neuroscientists also think there is a hierarchy of levels within the cortex (with primary sensory levels being the lowest/having the simplest representations or receptive field properties) and that these levels gradually get more complicated? So one question that would be useful to answer is how (and whether) prefrontal areas, for example, can modulate the properties or activity of other areas, sensory or motor.

Jye lyn Bold

The different brain regions involved in movement (volitional or not) appear to be described in the chapter based on a timeline of activation or chain of events rather than "higher-level" brain areas and "lower-level" ones. Do you think that the placement of events in the brain on the timeline is a more important distinction to consider with regard to exerting influence than categorizing brain regions as levels? Also, might the chapter eventually answer the question "How do higher-level brain areas exert influence over lower-level brain areas?" rather than the one posted in its title?

Replies to Follow-Up Questions

Mark Hallett

1. How is the brain constructed?
The brain might be thought of as a complex computer with many components. The basic building block of the brain is the neuron

analogous to a transistor in a computer. Each individual neuron has a very small influence. The brain is massively parallel at many levels, including at the level of the neuron. Indeed, many neurons are normally lost every day, and generally we don't notice. Many neurons, excitatory and inhibitory, together make a cortical column, cortical cluster, or, subcortically, a small nucleus, and this level might be thought of as analogous to a computer chip. Computer chips can make calculations for a particular purpose, and the cortical column can also. In the visual system a column might have output visual neurons that all respond to lines in a specific orientation. In the motor system, a column or cluster might have output neurons coding for movements of a body part in a specific direction, a movement fragment. Many columns together in a region make up an area (like a Brodmann area), and it is at this level where the component is large enough to have a defined function (areas are sometimes referred to as modules). In the computer analogy, the area or module would be like a circuit board. An area in the visual system would be able to process a whole visual image, and one in the motor system would be able to create commands to generate goal-directed movements. Each brain area is then connected to many other areas (the connected areas are called "networks"). Connections from one area to another are mediated by action potentials in the nerve axons going from the first area to the second and activating nerve cells in the second area by synaptic activity. In the analogy, we have now built the computer. Every area is not connected to every other area, but it is probable that after several steps every area might well be able to influence any other one. In the parlance of graph theory, the brain is a small-world network, which permits efficient and economic communication. It is the whole machine that is operative; the brain ordinarily functions as a whole. For any type of action, some areas are more important than others, and if an area of the brain is damaged, a specific malfunction might occur. Because of the massively parallel organization, if only one area is damaged, the malfunction might be minimal or quickly compensated.

2. Are the areas organized in a hierarchy, or are they all at the same level?

As just described, with all the areas connected and the brain working together, the areas certainly could be described as at the same level.

Moreover, most connections in the brain are reciprocal, so information and influence flow both ways. This influence could be the same in both directions. Thus, there would not be a hierarchy. Moreover, in the original answer, a specific multilayered hierarchy is described for making movement, but it is not necessary to use this exact hierarchy for making some types of movement. A different and limited part of the CNS can do the job. If you touch something very hot, your hand pulls away very quickly, even before you feel the pain. This movement is a flexor reflex mediated by a short connection between afferents and efferents in the spinal cord. Another example is the dual route from the visual system to the motor system; one goes via the perceptual system and the other runs straight to the motor system. If you "see" a snake, you might start running away before you "perceive" the snake. Thus, it is possible to consider all the parts of the CNS at the same level, and different parts can produce movement in different situations. (Although, of course, in the end, every connection must finally get to the muscle.)

3. Can modules in a hierarchy be thought of as going from simple to complex?

In the sensory systems, the sensory receptors identify elemental building blocks, and these get put together at successive levels to create objects. In the visual system, for example, information begins as points; many points become lines, lines become angles, angles become shapes. This makes sense as a hierarchy. And similarly, the operations in the motor system for many movements can be thought of in a hierarchy opposite to that in sensory systems, of complex to simple: a general goal (I want some chocolate), a general plan (I will get some in the kitchen), a specific plan (I will walk to the kitchen and take some from the third shelf), execution of a series of simple motor commands to execute the plan (a gait program to get to the kitchen, a reach-to-grasp program to obtain the chocolate, a hand-to-mouth movement, chew, swallow: goal accomplished!). These steps are likely carried out with information flow in a time sequence by the limbic network, the fronto-parietal network, and the sensorimotor network. Each network in turn activating the next network.

4. Do the higher levels' effects over the lower levels have to be in time order, or can it just be an influence of one level on another?

As just described, the influences from higher to lower levels are usually in a time sequence. The essence of influence is that the neuronal activity in the first level causes changes in the neuronal activity in the second level. This is easy to understand as a time sequence because we think that causes precede effects. However, it might well be that some influence can be without apparent time difference. As noted earlier, many connections in the CNS are reciprocal so that information can flow in both directions. Action potentials take time to go from one part of the brain to the other, but there are also action potentials taking the same amount of time going the other way. Each area then is affecting the other, and the two areas can appear to be acting synchronously. Nevertheless, a formal assessment of "effective connectivity" during a specific task can demonstrate a net influence in one direction. So it is possible to consider an influence without a time.lag, but this might be more apparent than real.

PART II, SECTION II

QUESTIONS ABOUT INTENTION

21

What are intentions and intentional actions?

Elisabeth Parés-Pujolràs and Patrick Haggard

While I am typing the answer to this question on my computer, I clumsily reached out for the glass in front of me and spilled water all over the desk. Did I mean to do that? On one level, I did want to take a sip, and hence I *intended* to reach out for the glass, since that is an action I could perform to achieve my goal. On another level, I did not mean to reach *like that*: I did not mean to move in such a way that I would tip the glass and spill the water. This example illustrates two core features of intentions and intentional action: first, that any given action can be described as intentional or unintentional depending on the level of description; second, that most of us typically know whether our actions are intentional or not, and can give detailed answers to the question "Why did you do *that*?" (Anscombe, 1957).

When people say they intend to do something, they normally mean that they have a certain goal and that they are planning to execute a certain action or sequence of actions to achieve it. One can say that intentional actions are those we do "for a reason."[1] This commonsense notion of intention has been conceptualized by philosophers as a hierarchy of intentions, including abstract goals ("distal intentions"; e.g., drinking), general action plans ("proximal intentions"; e.g., reaching for a nearby glass), and detailed motor commands ("motor intentions"; e.g., reaching with the right hand in such and such position, at such and such speed; Pacherie, 2008).[2] Although this is a conceptual distinction,

---

[1] See Chapter 15 by Sinnott-Armstrong on reasons.
[2] See Chapter 16 by Haggard & Parés-Pujolràs on the neural processes that produce actions.

Elisabeth Parés-Pujolràs and Patrick Haggard, *What are intentions and intentional actions?* In: *Free Will.*
Edited by: Uri Maoz and Walter Sinnott-Armstrong, Oxford University Press. © Oxford University Press
2022. DOI: 10.1093/oso/9780197572153.003.0021

some levels of the hierarchy have plausible neural correlates. Imaging studies have identified a hierarchical map of motor representations in the human cortex, ranging from rough general "patterns" of movement such as reaching and grasping to fine-grained tuning of the position of each individual phalanx of each finger at the lowest levels of abstraction (Grafton & Hamilton, 2007). Distal intentions, however, remain more controversial since no stable, context-independent representations have been found for abstract goals (Uithol, Burnston, & Haselager, 2014). In other words, while there are identifiable neurons that fire in recognizable patterns when a person intends to "grasp," there has been less success in identifying firing patterns for the more general intention "to eat an apple," still less for the more abstract intention "to get food." While neuroscientists have not identified stable correlates of abstract intentions, it is assumed that they are somehow encoded in the brain and that they drive action selection.[3] In classic neurocomputational models of action control, the brain "backward" calculates all possible ways to achieve a goal, and then selects the one that is best suited to the current context (Wolpert & Kawato, 1998). This process has been called "inverse modeling" and is well-accepted as a mechanism necessary for action selection.

This first approach to describing intentions provides a neuroscientific answer to the question about intentional action. Did I intend to do what I did when I tipped over the glass and spilled the water? The action was intentional to the extent that it followed from my distal intention to "drink" and my proximal plan to "reach" for the glass. Yet something went wrong in the specification of the motor command triggering my action, so I did *not* intend to execute *that* specific action. While the action in this example fails to be "fully" intentional, it would be labeled intentional at some higher levels of the hierarchy.[4] I can explain why I was reaching for the glass, and I identify the goal of drinking as my own. Often, these high-level, abstract aspects of an action are the most relevant ones for an agent's mental life, at least at the personal level of explanation. Thus, it seems that as long as we can identify our actions as part of a plan to reach a certain goal, we would

---

[3] See Chapter 22 by Haynes on evidence that intentions are represented in the brain.
[4] For more on hierarchy in the brain, see Chapter 20 by Hallett.

identify our actions as intentional. Even if the intentional actions fail, they will still be considered intentional. In fact, in many legal situations, the outcome of an action is only one of the elements considered in deciding responsibility. The intention an agent had when executing an action additionally plays a major role.

This brings us to the second main relevant aspect of intentional action: the fact that people (except for some interesting pathological cases) typically *know* whether their actions are intentional or not, and in what sense they are intentional.[5] While it is relatively easy to investigate and find the neural correlates of specific motor actions, identifying the neural basis of the "feeling" that is typically associated with voluntary actions is harder. As we have seen, we do not have robust neural markers of people's more abstract intentions. However, it is possible to determine the extent to which someone feels responsible for the *outcomes* of their actions in experimentally controlled settings. A phenomenon known as "intentional binding" describes the fact that people tend to perceive the outcomes of intended actions as closer in time to the action execution than outcomes of unintended actions that were externally triggered rather than endogenously initiated (Haggard, Clark, & Kalogeras, 2002). The magnitude of the "binding" is considered a marker of the extent to which people feel responsible for their actions, or, in other words, the extent to which they "intended" them in the relevant sense. Interestingly, it has been shown that people executing actions under coercion have reduced intentional binding, and that the outcomes of their actions are processed differently by their brains (Caspar, Christensen, Cleeremans, & Haggard, 2016). Thus, even though we do not know precisely how abstract intentions are encoded in the brain, we can track the experience people have of their own intentionality by means of indirect behavioral measures. Of course, the diagnostic value of such measures can be questioned (e.g., Suzuki, Lush, Seth, & Roseboom, 2019).

In sum, intentional actions can be defined as actions that follow from an agent's goals and are identified as such by the agent herself. While neuroscience has a fairly good idea of the neural representations

---

[5] See Yaffe's replies in Chapter 1 on intention.

that underlie the ability to execute a specific action,[6] determining how abstract intentions drive the selection and activation of detailed motor representations remains a challenge. In turn, the combination of behavioral markers and imaging techniques offers interesting insight into the neural basis of the distinct feeling of intentionality and outcome-directedness that accompanies most voluntary actions.

Follow-Up Questions

Walter Sinnott-Armstrong

You conclude with "intentional actions can be defined as actions that follow from an agent's goals and are identified as such by the agent herself." In contrast, philosopher Gideon Yaffe says in his chapter in this volume, "To have an intention is to have a thought about a future action, where the thought has certain distinctive effects on behavior and reasoning and is subject to distinctive rational requirements."[7]

(a) How is your definition related to Yaffe's? Do the kinds of intentions that you discuss—distal, proximate, and motor intentions—also fit Yaffe's account?
(b) Yaffe requires causation or "distinctive effects" as well as a special relation to rationality. Are these additional requirements compatible with your account of intentional actions?

More generally, are philosophers and neuroscientists talking about different things when they discuss intentions?

Adina Roskies

It seems that the account of knowing whether one intends something is compatible with a purely backward-looking and possibly inferential

---

[6] See Chapter 16 by Haggard & Parés-Pujolràs on the neural processes that produce actions and Chapter 20 by Hallett on levels and control in the brain.
[7] See Chapter 1 by Yaffe on intention.

account of what that knowledge consists of. Does that raise the possibility that one can act intentionally without being conscious at the time of action of one's intentions?

Replies to Follow-Up Questions

Elisabeth Parés-Pujolràs and Patrick Haggard

Our description of intentional action left two crucial questions open. First, how do intentions *cause* actions? Second, how does consciousness relate to intentionality? We discussed in this chapter that intentions can be described at different levels of abstraction. Crucially, the ability to achieve the more abstract goals depends on an interaction between those higher-order representations and the lower-level ones. Which form does this interaction take? One way to think about the role of intentions in determining behavior is as "structuring causes" (Dretske, 1988). This seems similar to "distinctive effects" of the intention on behavior. Presumably, the most distinctive effect of intentions is to make some corresponding action(s) likely to occur, or at least subjectively likely to occur. Goal representations configure how lower-level decisions are made, and therefore ultimately determine which actions are triggered (Goschke, 2013). Imagine, for example, that you are looking for a place to eat tonight. At a neural level, simple decisions such as which restaurant to pick are often conceived in terms of a competition between action representations. Several factors will influence this competition, and hence your subsequent choice. If available competing options are subjectively equivalent (e.g., if you are equally inclined toward a Greek and an Italian restaurant when you need to decide), then the decision may depend on random variations either in the environment (e.g., some Greek lettering you just happened to see reminds you of your Greek vacation and biases toward the Greek restaurant) or in neural activity (e.g., the memory of that Greek vacation might happen to be temporarily more salient while making the decision). Instead, if you have reasons to prefer one over the other, the neural representation of those reasons can influence the way your decision goes. For example, if you are very hungry, you may prioritize nearby restaurants rather than other good ones that are further away. Alternatively, if you are a vegetarian you will probably prefer restaurants that offer vegetarian options to nearby restaurants that do not, even if you are hungry.[8] By sending excitatory or inhibitory input to the neural populations that represent alternative options, neural representations of goals can bias decision-making processes.

As we argued earlier, it is unclear which brain area holds these types of goal representations. However, the prefrontal cortex (PFC) has been suggested as a candidate area. It sends output to brain areas involved in decision-making, and its neural activity is compatible with task-encoding and long-term goal representations. But we often lack clear criteria for interpreting exactly what any particular neural activity represents. When we record from a particular neuron or measure a particular pattern of blood flow, how can we say whether we are dealing with a plan, a goal, a thought, an intention? We can't. The most we can hope to do is to compare one representation to another and infer the position of one representation relative to another along the route that allows "the thought to have certain distinctive effects on behavior and reasoning." Both philosophy and neuroscience view intentional action as a causal pathway, with a clear direction from thought to action, and thus fitting the world to its representation in our mind, a world-to-mind fit.[9] In our view, this pathway is informational as well as causal. The progression from thought to action involves decisions, choices, and specifications, as our restaurant example shows. Intentional action can therefore be considered a kind of information generation.

As we have seen, the ontology and meaning of any particular stage along that pathway can be obscure in philosophy, and often in neuroscience too. Sinnott-Armstrong asks, "Are philosophers and neuroscientists talking about different things when they discuss intentions?" Possibly not. Yet it is often experimentally difficult to confirm whether a specific pattern of neural activity meets all the criteria that philosophical accounts require of intentions. One-to-one correspondence between philosophical and neuroscientific accounts may be unlikely. However, focusing on the relations *between* constructs

[8] For more on the role of reasons in decision-making, see Chapter 19 by Bold et al.
[9] See Chapter 1 by Yaffe on intention.

may be more promising. Two theoretical constructs in the philosophy of intentional action may differ in a clear and explicable way. Similarly, neuroscientists may be able to measure two distinct processes in the preparation of voluntary action. We may then ask whether the *difference* between the theoretical constructs corresponds to the *difference* between the neural representations. For example, the philosophical distinction between distal and proximal intentions seems comparable to the neurophysiological distinction between the general readiness potential (which is typically taken to reflect a general state of movement preparation) and the lateralized readiness potential (which reflects preparation to make a movement with either the right- or the left-hand side of the body).

What role does *conscious* thought play in this type of causality? Achieving a goal typically involves a series of actions. Each of those actions, as well as the goal itself, may or may not be present in consciousness at any given point in time. While our definition of intentions as neural representations of varying levels of abstraction does not imply that *all* or *any* of those are conscious, a definition of intention as "thoughts about future actions" presumably does require conscious experience of those thoughts.

Our account leaves open the possibility that when asked "Why did you do that?" people confabulate. People may offer inferences or plausible explanations that could account for the observed behavior but did *not* precede it, and hence could not have caused it. In the examples we offered, one is most likely conscious of the reasons that motivate one's choices (i.e., being hungry or being a vegetarian). However, in other cases one may not be. It is certainly possible that people retrospectively infer causes for their actions, and there is broad evidence that people indeed do this, at least sometimes. However, this does not mean that those same actions about which we can confabulate were not intentional in the first place. Subsequent confabulation does not imply lack of a relevant and conscious intention at the time of the action. For example, the original intention might have gone through a process of Orwellian revision (Dennett, 1991), so that another explanation for the action is now given. In fact, the most common marker of voluntary action might not be a strong conscious "fiat" but rather a certain tacit "awareness of consent" (Ach, 1910/2006). We often absent-mindedly execute voluntary actions that do not feature prominently in our conscious stream of thoughts, and yet they follow from an internal state that we implicitly recognize without necessarily having a vivid experience of it. Consider, for example, the all-too-frequent scenario where someone wanders into a room and then finds herself thinking, "Wait, why did I come here in the first place?" In those cases, she typically does not doubt that she walked there voluntarily and that she had some intention motivating her to do that—simply she can no longer access the original intention. Yet even in these cases in which we fail to recall the reasons behind our actions and where there may not even have been any conscious thought "triggering" them, its voluntary *feel* remains intact. In that sense, our position is compatible with the idea that one can act intentionally without an explicit, conscious reason.

22

What evidence is there that intentions are represented in the brain?

John-Dylan Haynes

Various signals have been observed that reflect different aspects of encoding of intentions,[1] although the exact nature of intention encoding in neural signals is still only poorly understood. Importantly, not much is known about how the degree of commitment to an intention affects its neural representation.

Nature of the Intention

(A) Simple Motor Intentions

One way to assess the encoding of motor intentions is to observe animals that have been instructed ("cued") to perform one of several possible movements and that are waiting for a cue to perform this movement (say, pulling a lever in one of several possible directions). The activity of single cells in the primary motor cortex and in the premotor cortex is elevated during this preparation period. The activity exhibits a movement-specific tuning such that different cells increase their firing rates for different movements. Similarly, human neuroimaging studies have shown that when participants are waiting to perform a specific movement, the specific movement they will carry out can be decoded from delay-period signals in motor cortex. (B) Task sets: For simple motor intentions the exact movement is pre-specified

---

[1] For more on intentions, see Chapter 1 by Yaffe and Chapter 21 by Parés-Pujolràs & Haggard.

John-Dylan Haynes, *What evidence is there that intentions are represented in the brain?* In: *Free Will*. Edited by: Uri Maoz and Walter Sinnott-Armstrong, Oxford University Press. © Oxford University Press 2022. DOI: 10.1093/oso/9780197572153.003.0022

at the time of the intention. In many cases, however, a correct move-ment depends on properties of a stimulus that has not yet been presented, say, when waiting to press a left or right button depending on whether a picture that is about to appear on the screen is a face or a house. In this case, the preparation consists of establishing an abstract rule that maps specific stimuli to specific movements without the ability to pre-specify the exact motor plans. This can occur in a simple fashion, as is the case in simple stimulus-response mappings (e.g., when you see a house, press the left button); alternatively, such rules can be more complex (when the parity of the number you are about to see is odd, then press the right button). When animals are cued to perform one of several possible complex tasks, the firing rate of cells in lateral prefrontal cortex is modulated by which task the animal is currently preparing to perform. For example, the firing rates of individual cells differ depending on whether the animal is preparing to perform a task that involves analyzing objects with respect to their identity, their spatial location, or a combination of both. The tuning of individual prefrontal cells to specific tasks does not follow a "sparse" or "labeled line" code, but instead exhibits mixed selectivity, where each cell participates in encoding various intentions, and only the pattern of activity across the population can be used to reveal the exact current intention. Human neuroimaging studies have shown that task sets can be decoded from fMRI signals in various regions, including medial and lateral prefrontal cortex and parietal cortex. Some have argued that such neural codes of intentions might be highly context-selective and thus might not reflect a single underlying representation.

Cued vs. self-chosen intentions: An important question is whether task-selective neural activity depends on the way the intention to perform a task comes about—specifically whether the person chooses themselves what to do, or whether the task is instructed by the experimental conditions. In the motor domain, both cued and self-chosen, spontaneous movements activate primary motor cortex and supplementary motor area. There is surprisingly little difference in brain signals between these two conditions. During spontaneous movements the signal begins earlier in supplementary motor area (SMA) (relative to primary motor cortex, M1) than it does for cued movements. In human neuroimaging experiments, the pre-supplementary motor area (preSMA)[2] and dorsal anterior cingulate cortex (dACC) have sometimes been observed to be more activated for freely chosen than for cued intentions. However, studies using multivariate decoding have suggested that the neural representation of cued and self-chosen intentions is very similar across a wide range of intention-related brain regions.

Proximal vs. distal intentions: An important distinction is whether intentions relate to an action that is performed right now (I am stirring the soup now), to an action that is being prepared and is the next to be performed (I am standing in front of the stove waiting for the soup to boil, and then I will stir it), or actions that are going to be performed only after a delay during which participants are engaged in other activity (I am giving a lecture now, but later on I will cook dinner, which will involve my stirring the soup). Neural representations have been found for all three types of intentions with somewhat dissociable brain regions. For example, when a different task has to be performed first, before engaging in the relevant target action, then the action is encoded in more anterior regions of prefrontal cortex. This region is involved in prospective memory, because the ability to maintain intentions across delay periods deteriorates with lesions in this region.

Motivation: An important property of intentions is that we are typically motivated to carry them out. Thus, the question arises how urges and desires[3] affect the encoding of intentions. Waiting for naturally and spontaneously occurring urges and desires is difficult in structured experimental settings. So the standard approach in neurocognitive experiments is to vary the reward for correctly performing an action and with it the motivation to carry out that action. Regions of the basal ganglia and of orbitofrontal cortex have been shown to encode reward signals during task preparation. Orbitofrontal and parietal brain regions have been shown to encode links between actions and rewards. The encoding of task sets is also strengthened when participants are more motivated to perform a task.

---

[2] The pre-supplementary motor area is the anterior part of supplementary motor area (SMA). The SMA is featured in Figure 2b in the Brain Maps section.

[3] See Chapter 23 by Hopkins & Maoz on the neural correlates of beliefs and desires.

Intention-predictive brain signals: There is also activity in the human brain that *precedes* the conscious formation of an intention: a negative-going EEG signal called the "readiness potential" (RP) is observed over motor-related brain regions in the final few hundred ms preceding an urge to perform simple movements.[4] Single neurons in the supplementary motor area of patients show a similar progressive increase in activity prior to the formation of an intention. In fMRI, brain signals in prefrontal and in parietal cortex can be used to decode and predict to some degree which intention a person will form several seconds later. The limited accuracy could mean either that the brain only partially biases the decision ahead of time or that the limitations of fMRI preclude us from measuring the full predictive information that would be contained in invasively recorded neural signals.[5] Either way, the combination of EEG recordings and real-time brain-state decoding suggests that such intention-predictive brain signals are not deterministic because a decision for a specific action can still be revoked when required so by a task ("veto").[6]

Follow-Up Questions

Walter Sinnott-Armstrong

You usefully discuss simple motor intentions and distinguish cued from self-chosen intentions and proximal from distal intentions. Do all of these kinds of intentions fit within the definitions of intentions in other chapters of this volume? Parés-Pujolràs and Haggard conclude, "[I]ntentional actions can be defined as actions that follow from an agent's goals and are identified as such by the agent herself." In his chapter on intention, Yaffe writes, "To have an intention is to

---

[4] For different angles on the RP, see Chapter 9 by Roskies, Chapter 16 by Haggard & Parés-Pujolràs, and Chapter 27 by Hallett & Schurger.

[5] For more on fMRI and neuroimaging in general, see Chapter 27 by Hallett & Schurger.

[6] The author of this chapter recommends Haggard (2019); Haynes (2011); Schultze-Kraft et al. (2016); Schurger and Uithol (2015) for further reading.

have a thought about a future action, where the thought has certain distinctive effects on behavior and reasoning and is subject to distinctive rational requirements." Would all of the intentions that you discuss count as intentions under their definitions? If not, is there evidence that intentions of the kinds that they have in mind are also represented in the brain?

Aaron Schurger and Ned Block

How can we distinguish (proximal) "intention" from "preparation"? Is there a difference? Conceptually? Empirically? In general, what is a good operational definition of "intention" that distinguishes it from mere "precursors" or "antecedents" or "preparatory activity" or "memory"? Moreover, intentions are often present alongside the suppression of other intentions, the latter of which requires executive function. How do you therefore separate intention from executive function?

Adina Roskies

You write, "An important property of intentions is that we are typically motivated to carry them out." Are we only "typically motivated," or is this motivation somehow constitutive of intention? (Could we have an intention to do something without any motivation whatsoever?)

Replies to Follow-Up Questions

John-Dylan Haynes

As Sinnott-Armstrong points out, other chapters in this volume have sought to explicitly define the term "intention" with somewhat different results and different emphases. For example, in their chapter, Parés-Pujolràs and Haggard consider the fact that an action subserves a person's self-stated goal as a key criterion for an intentional action.[7] Some (but not all) of the studies reported earlier involve cued tasks where participants are told by the experimental instructions what to do. This raises the question whether such cued (one might also say "requested") tasks can be considered true goals. Let's take an example. My wife asks me to go to the supermarket and buy something, say, a packet of shrimp. I do not like or eat shrimp, so I am doing her a favor by buying it for her. If you were to ask me on my way to the supermarket what my intention is, I would say, "I am intending to buy some shrimp for my wife." I would clearly consider this an intention. When asked "Why are you buying the shrimp?" I would happily give a reason[8] ("to make my wife happy"). So even if it was originally someone else who requested that I pursue the goal and it originally reflects their interests and not my own, I have now taken it and made it my own. Take a different example. Say I participate in an experiment and I am asked to press the left button if I see a face. When asked if I intend to press the left button when I see a face, I would answer "Yes, because that is what the experimenter asked me to do." There is no reason that a cued task cannot be turned into a true intention. This does not mean that I will not pursue intentions more vigorously when they are based on my own internal desires, just that both can be intentions.

Another important question, by Aaron Schurger and Ned Block, is whether it is possible to distinguish between neural representations of intentions and those of other correlated cognitive factors. Let's start with separating intentions from preparation to move. When I am asked to do something (say, indicate which object is visible on the screen), I will also memorize the instruction (the cue given on the screen) and prepare for the action (e.g., pre-activate the motor representations in the brain of the hands needed to indicate the result). Because the intention and the motor preparation are correlated, it can be tricky to dissociate them. However, there are ways to design experiments that provide at least a plausible separation. For example, an intention is often mentally invariant to the different ways of executing a corresponding action (e.g., a button press with the left hand, a pedal press with the right foot). In such cases, whether a representation is an intention or a preparation can be tested by assessing

---

[7] See Chapter 21 by Parés-Pujolràs & Haggard on intentional actions.
[8] See Chapter 15 by Sinnott-Armstrong on reasons.

its invariance to different forms of execution. Say that I plan to add numbers; does the neural representation that was identified pertain to a specific number I have to calculate or to a specific hand with which I need to report the result? Or is it invariant to the details of the action? We have tested this and found that there is indeed a degree of (but not a complete) invariance of such representations. Furthermore, it is common practice to use multiple cues to instruct the same task (e.g., a red circle and a blue square could both instruct the subject to press the left button with their left hand). This is in order to test whether task representations are invariant to the memorized visual properties of the cue by training a classifier on one set of cues and testing on another. It is important to build such testing procedures into the experimental design of a study. Finally, there is some evidence that the representation of one intention is somewhat invariant to the other intentions that could be considered, suggesting that it is not mere suppression of competing plans. For example, the spatial brain pattern encoding an intention is somewhat invariant with respect to the set of other, unchosen intentions that are involved in a study. Also, we have directly shown the encoding of intentions in prefrontal cortex while competing plans are held simultaneously.

A second way to assess whether a representation is indeed an intention (rather than a preparation or a memory) would be to test whether it is resistant to change. Say I will be shown a picture of a house or of a face on the screen, and I intend to report the house by pressing a button with my index finger and the face by pressing a different button with my middle finger, both of my right hand. For simplification we can consider two representations here: the representation of the overall plan as described and the pre-activation (i.e., preparation) of the two finger representations in motor cortex. Let us speculate: What if I were to use Transcranial Magnetic Stimulation (TMS) to disrupt the preparation in motor cortex? Given that the intention has a "world-to-mind" direction of fit, presumably the activity of the motor cortex that reflects the correct preparation would be restored by the circuits reading out the intention and preparing the execution. If, however, I were to disrupt processing in the intention representation in the brain, I could change the intention itself so that it might not be restored. I am not aware of any studies directly testing this, but it shows that in principle one can conceive of ways to test for different representational roles.

Please note that the resistance to change can come in varying degrees, thus also providing a potential handle on the role of motivation. One would assume that a person with a high motivation to perform a task (say, because they receive a high reward) would try to maintain their intention against distraction better than if they had only a low motivation. This has indeed been found. In this respect, Adina Roskies's question about whether or not motivation is a necessary ingredient in intentions is also very interesting. Perhaps one could conceive of intentions that occur in the absence of motivation? Here it might help to consider an extreme case, e.g., a movement that is directly opposite to the motivational state, say, an unwanted nervous tic. This would presumably not be considered an intention. Also, when one is indifferent to the action, this would presumably not be considered intentional. Thus, a motive (or desire) to perform an action seems to be necessary for the formation of an intention. Nonetheless, encoding of the different aspects of intentions are dissociable in neural terms. For example, the motivation, the content of the intention, and the execution conditions (e.g., when I want to perform the intention) appear to be encoded in separable brain regions.

Finally, one can postulate that intentions—whether simple, motor ones or more abstract—are subject to constraints on rational consistency. As an example of a motor intention, think about a cutter machine that requires the operator to press two buttons that are separated so far that it is impossible to press them with one hand only. The operator has to use both hands as a safety measure to avoid a hand getting caught in the machine. Here, the operator might have the desire to use just one hand, simply because it is easier. But at the level of motor intentions, actions need to be rationally consistent with how the world lets me achieve a desired effect. The neural mechanisms of how motor intentions are adapted to multiple constraints are currently not well understood.

23

What is known about the neural correlates of specific beliefs and desires that inform human choices?

Amber Hopkins and Uri Maoz

Some theories of free will, especially of the compatibilist bent, highlight the importance of acting in accordance with one's beliefs and desires.[1] So, one manner by which neuroscience could contribute to the free-will debate is by clarifying what is known about how beliefs and desires play out in the brain: which regions are involved, when activations take place, and their impact on areas implicated in the planning and execution of action. This understanding could then constrain philosophical theories of free will. However, it is important to acknowledge that there is little neuroscientific research on beliefs and desires that abides by the precise manner in which these concepts are defined and analyzed in the relevant philosophical discourse. While we do not have neuroscientific evidence to wholly elucidate beliefs, desires, their interactions, and their exact roles in voluntary action, there is nevertheless interesting neuroscientific work to report.

Neuroscientific research on the neural correlates of beliefs and desires highlights many regions for each. Areas associated with belief include the prefrontal cortex (PFC), more specifically the ventromedial prefrontal cortex (vmPFC), ventral prefrontal cortex (vPFC), and medial prefrontal cortex (mPFC). Other associated brain regions include the anterior cingulate cortex (ACC), inferior frontal gyrus (IFG), temporal parietal junction (TPJ), ventral striatum, ventral tegmental area

---

[1] For more on compatibilism, see Chapter 4 by Sinnott-Armstrong on freedom, Chapter 5 by O'Connor on free will, and Chapter 6 by O'Connor on determinism.

Amber Hopkins and Uri Maoz, *What is known about the neural correlates of specific beliefs and desires that inform human choices?* In: *Free Will.* Edited by: Uri Maoz and Walter Sinnott-Armstrong, Oxford University Press. © Oxford University Press 2022. DOI: 10.1093/oso/9780197572153.003.0023

(VTA), substantia nigra pars compacta (SNc),[2] and the mesocortical dopamine pathway. Areas potentially implicated in desire include the vmPFC, orbitofrontal cortex (OFC), cingulate cortex, ventral striatum, nucleus accumbens,[3] the limbic system, amygdala, VTA, SNc, and the mesolimbic dopamine pathway.

How beliefs and desires interact in the brain to inform choice and guide action may be interesting to explore here. The PFC, and especially the vmPFC, and its dopaminergic innervations are described as a possible neural correlate of the cooperation of beliefs and desires in human choices.

The Ventromedial Prefrontal Cortex

The PFC is associated with cognitive control (or our ability to align our actions with our goals), planning, decision-making, reasoning, working memory, sensory integration, contextual appraisal based on previous experience, and regulation of movement. Thus, the PFC likely plays a part in our capacity for abstract or complex mental states, along with facilitating their contributions to choice and action. However, at a more precise spatial scale, the vmPFC appears to be especially relevant here.

The vmPFC is often tied to value-based, or preference-driven, decision-making, along with emotional processing, self-perception, and social cognition. This region is active when valuating or assessing the positive and subjective value of choice alternatives, rewards, or emotional stimuli (Bartra, McGuire, & Kable, 2013). So it seems that when we are prompted with a choice, determining which alternative is most desirable to us, or acting in accordance with our own desires, could recruit the vmPFC.

---

[2] The ventral striatum is the ventral part of the striatum, and the substantia nigra pars compacta is part of the substantia nigra. The striatum and the substantia nigra are parts of the basal ganglia, the latter being featured in Figures 1b and 3 in the Brain Maps section. In contrast, the ventral tegmental area is located in the midbrain, which is featured in Figures 1b, 1c, and 3 in the Brain Maps section.

[3] The nucleus accumbens is part of the ventral striatum, the striatum being one of the nuclei of the basal ganglia. The basal ganglia are featured in Figures 1b and 3 in the Brain Maps section.

In addition to the vmPFC's involvement in valuation, this region is also associated with forming beliefs, making use of beliefs, and emotion-laden reasoning. The vmPFC has been tied to valence-guided belief formation—the selective formation of beliefs regarding oneself that are favorable or desirable—for example, in optimism bias (Kuzmanovic, Rigoux, & Tittgemeyer, 2018). Additionally, when this region is active, individuals are more likely to make choices in accordance with their belief-biases (Goel & Dolan, 2003). What is more, the vmPFC is commonly activated when engaged in both emotional and moral reasoning, or when making moral judgments (Blanchette, 2013).

Finally, the vmPFC is anatomically connected to two regions that are important for action. The first is the dorsolateral prefrontal cortex (dlPFC), which is tied to cognitive or executive control of behavior, which may include action initiation, and influences the premotor cortex. The second are the basal ganglia, which have been implicated in the control of voluntary movements.

Given its associations with valuation and desire, belief and reasoning, and its anatomical connections to motor regions, the vmPFC appears to be involved in modulating the effects of emotion and desire on cognition, choice, and action (Goel & Dolan, 2003). So there is considerable evidence that the vmPFC is involved in the combination of beliefs and desires in informing, or guiding, our choices.

Dopaminergic Innervations

The neurotransmitter dopamine and reward are regularly discussed in tandem. Dopamine is implicated in both wanting, a motivational state, and reward, which can be understood as the attractive and motivational property of an external source. Dopamine induces pleasure when a rewarding outcome is acquired, and the absence of this neurotransmitter incentivizes us to act in ways we have learned will yield it.

The role of dopamine in sustaining our interest in rewards or reward-paired stimuli, and so in desire, is well-established. The mesolimbic pathway is a dopaminergic pathway that connects the VTA (a dopamine-producing region associated with reward cognition, reinforcement learning, wanting, and liking) to the nucleus accumbens in the ventral striatum, which seems to be involved in reward processing and motivation, but also decision-making.

However, dopamine seems to go well beyond reward-related functionality; it has also been tied to cognitive flexibility, gating sensory input, facilitating the flow of sensory and contextual information to and within subregions of the PFC, and, crucially, updating prefrontal representations of, and so perhaps beliefs about, one's surroundings (Ott & Nieder, 2019). The mesocortical pathway, a dopaminergic pathway that constructs streams, or "loops," of communication between the VTA and the PFC, is especially linked to belief formation. Abnormal dopaminergic activity along this pathway is associated with aberrant belief formation, such as delusions in psychosis.

Furthermore, dopamine appears to be vital for controlled movement.[4] The nigrostriatal pathway, a dopaminergic pathway connecting the dopamine-generating SNc to the dorsal striatum[5] (both subcortical nuclei in the basal ganglia), is associated with motor regulation. The imperative role dopamine plays in movement is revealed when dopaminergic neurons in the SNc degenerate, thus reducing dopaminergic input to the striatum and resulting in motor control impairments, such as some of the symptoms of Parkinson's disease.

Given its well-established role in reward, dopamine is likely related to desire, yielding reward, pleasure, and wanting—a motivational pull toward certain choices and actions. However, dopamine and dopaminergic pathways have also been connected to updating prefrontal representations about the external world, to supporting belief formation, and to facilitating controlled movement. Thus, dopamine and dopaminergic pathways seem to be involved in both beliefs and desires and likely mediate their impact on choice and action.

---

[4] See also Chapter 20 by Hallett on levels and control in the brain.
[5] The dorsal striatum is the dorsal part of the striatum, which is one of the nuclei of the basal ganglia. The basal ganglia are featured in Figures 1b and 3 in the Brain Maps section.

Concluding Remarks

The vmPFC and dopaminergic pathways in the brain are leading candidates to be parts of the neural networks implicated in beliefs and desires that inform human choices and actions. The vmPFC is associated with valuation, preference-driven decision-making, choosing in accordance with beliefs, forming beliefs, and reasoning. It is anatomically connected to the dlPFC and basal ganglia, which are related to behavioral control and the initiation of voluntary action. This connection could therefore pave the way for beliefs and desires to launch and guide action.

Dopamine's role in reward, pleasure, and wanting—and thus in desire—is fairly well-established. However, dopaminergic innervations that connect dopamine-rich midbrain regions to subregions of the PFC, likely including the vmPFC, have been linked to our capacity to cognitively represent our surroundings and form beliefs about the external world. Moreover, dopamine is vital for motor control. So, independently or together, the vmPFC and dopaminergic pathways in the brain are relevant for discussions about the neural bases of beliefs and desires influencing choice and action.

However, despite the above, much more investigation is needed to better unravel the neural underpinnings of specific beliefs and desires, their interactions, and their specific impacts on regions related to movement, especially in accordance with these concepts in philosophy. Given the importance of such work in elucidating volition, we hope for more neuroscientific studies rooted in the crucial conceptual distinctions that have been delineated about beliefs and desires in the philosophical literature.

Follow-Up Questions

Mark Hallett and Timothy O'Connor

You note that many areas throughout the brain have been implicated in representing beliefs and desires. But you then mainly focus on the PFC. Would it be fair to say that the PFC is likely only one node in a

widely distributed network that would actually be the proper correlate of any specific belief? In other words, on current evidence, might
the intentional and informational content of beliefs and desires be distributed, rather than localized, while the vmPFC plays a crucial role in
connecting these distinct contents in informing choices?

Jonathan Hall

Would you agree that your description of the vmPFC corresponds
nicely to the reasons-responsive mechanism that is considered central to some compatibilist theories of free will (Fischer & Ravizza,
1998)? That, of course, does not imply that any animal with a vmPFC
has free will. Probing further about distinctively "human choices,"
what are the differences between human vmPFC and the equivalent
areas in primates? There are two potentially overlapping explanations
for what is distinctive about human reasons-responsiveness. One
focuses on the complexity associated with the wide variety of kinds
of reasons to which humans are responsive, consistent with a variety
of goals. The second focuses on the kind of reasons to which humans
are responsive—e.g., humans are uniquely responsive to deliberative
judgements that aim at "the True and the Good" (Wolf, 1990).

Walter Sinnott-Armstrong

Some philosophers try to analyze intention and will in terms of
beliefs and desires, but their analyses are controversial. How do you
understand the relations of beliefs and desires to intention and will
(as discussed in other chapters in this volume)? Can I intend to perform an action without believing that it will fulfill my desires better
than any alternative? Can I believe that an action will fulfill my
desires more than any other available option but still not intend or
will to do it? In particular, what if my different desires and beliefs
push equally strongly in opposing directions—like Buridan's ass? Am
I still able to intend and will (or choose) to move in one direction instead of the other?

Replies to Follow-Up Questions

Amber Hopkins and Uri Maoz

The relations between beliefs, desires, and intentions are—just as indicated in the question posed by Walter Sinnott-Armstrong—controversial and unsettled among philosophers. Beliefs and desires can both be understood as propositional attitudes, which connect agents to propositions, but also as reasons for action. However, beliefs, desires, and intentions can differ in several ways.[6] Certainly, the right kind of neuroscientific evidence could shed light on how beliefs and desires interact, how beliefs and desires relate to intentions, and, of course, how beliefs and desires influence choice and voluntary action. Unfortunately, we have yet to acquire the neuroscientific evidence to clarify whether or not we can intend without believing, believe without intending, or believe and act in ways that conflict with our desires. In the philosophical discourse, beliefs and desires are distinct and clearly delineated. We do not yet know to what extent this holds for their direct neural representation (if any). There is therefore exciting and important work to be done in this regard.

Regarding the relation between the will, beliefs, and desires, we appeal to an interpretation of "the will" eloquently expressed by Pamela Hieronymi in this volume.[7] Hieronymi describes one account of the will as a "collection of more or less ordinary, interacting aspects of the person's psychology (their cares, concerns, beliefs, desires, commitments, fears, etc.) that generates intentional, or voluntary . . . activity." Each of us is shaped by our experiences and unique in our beliefs and desires, so regions or networks implicated in belief and desire likely vary substantially between people. Inasmuch as our beliefs and desires are rooted in our brains, "the will" could then be, at least in part, constituted by the networks implicated in our individual collections of beliefs and desires.[8]

---

[6] Compare with Chapter 1 by Yaffe on intention.
[7] See Chapter 2 by Hieronymi on the will.
[8] See Chapter 17 by Kreiman on the neural basis of the will.

It can feel intuitive to start from states or experiences, in this case beliefs and desires, and then ask "Where in the brain does this happen?" with the hope that identifying genuine neural correlates will paint these complex philosophical dilemmas in a more black-and-white manner and perhaps settle them. While there is some specialization among brain areas, there is also shared functionality between brain regions and great inter-subject variability.[9] It thus seems unlikely that there would be a spatially precise region in the brain that is the genuine neural correlate of a specific belief or desire and more likely that beliefs and desires are supported by a complex network of many interconnected brain regions, as suggested in the question posed by Mark Hallett and Timothy O'Connor. The PFC in general, the vmPFC, and dopaminergic pathways are highlighted in this chapter; but it is more likely that they are but nodes in a vast network that represents specific beliefs or desires. Thus, their content would be more distributed than localized.

As pointed out by Jonathan Hall, the involvement of the vmPFC seems compatible with the conception of beliefs and desires as reasons for action or our capacity to be reasons-responsive, which is important in the free-will debate. The vmPFC is associated with processing goal-relevant information, preference-driven or value-based decision-making, choosing in a belief-biased manner, reasoning with emotional content, and making moral judgments (Blanchette, 2013). So the vmPFC appears to be a strong candidate to at least be part of a network that supports reasons-responsiveness, especially emotion-laden reasons. Alternatively, the cooperative activation of and connectivity between the vmPFC and the dlPFC could permit human responsiveness to a wider variety of reasons. The dlPFC has been tied to reasoning with emotionally neutral content and engaging in self-control, which involves being responsive to reasons *not* to choose some alternative or *not* to go through with some act (Hare, Camerer, & Rangel, 2009).

To further elucidate the vmPFC's role in reason-responsiveness, Hall suggests comparing the vmPFC or equivalent regions in primates to humans. Primate vmPFC appears to be involved in processes similar

---

[9] For more, see Chapter 27 by Hallett & Schurger on neuroimaging.

to those in humans, namely, encoding value or reward and flexible decision-making (Hiser & Koenigs, 2018). However, if humans are responsive to more complex or abstract reasons than other primates, and the vmPFC is crucial for this capacity, then human vmPFC may have some enhanced quality. A recent study assessed the relation between vmPFC volume and feeding ecology in five primate species, including humans. The authors predicted that the vmPFC would be more pronounced in primates with a rich diet and who engage in complex foraging strategies that require flexibility and planning. Interestingly, the authors report a positive relation between diet diversity and overall vmPFC volume as well as relative vmPFC volume, or the ratio of vmPFC volume to whole-hemisphere volume (Louail, Gilissen, Prat, Garcia, & Bouret, 2019). On another note, within humans, we associate a greater capacity for self-control and responsiveness to complex reasons with maturing. So it could be interesting to track the human vmPFC developmentally. One study suggests that functional coupling between the vmPFC and the dlPFC increases as the capacity for behavioral control or resisting immediate temptation develops from childhood to adolescence (Steinbeis, Haushofer, Fehr, & Singer, 2014). We could therefore speculate that, as we mature and respond to more complex reasons in accordance with a wider variety of goals, the interaction between the vmPFC and dlPFC increases.

PART II, SECTION III

QUESTIONS ABOUT CONSCIOUSNESS

24

How can we determine whether or not an agent is conscious of a bit of information relevant to an action?

Liad Mudrik and Aaron Schurger

The posed question is complicated and is actually a sub-question of a long sought-after, thus far unanswered, principle philosophical question: *How can we determine if an agent is conscious of a bit of information?*, which in turn is a sub-question of *How can we determine if an agent is conscious?* The latter is also known in philosophy as "the problem of other minds" and has been a matter of ongoing interest and discussion. Thus, this question might just as well have been presented to philosophers by neuroscientists.

That said, let us examine how the neuroscientific community—or that of the cognitive sciences in general—typically addresses this question. As always in science, this is done via some operational definition (a translation of a theoretical construct one wishes to study into empirical terms that can be manipulated and measured); the theoretical construct *being conscious/being conscious of a bit of information/being conscious of a bit of information related to action* is translated to observable phenomena that can be measured by researchers and serve as a working definition.

Let us therefore unpack our answers to these questions from the more general to the more specific. *Being conscious* is commonly studied in the context of "state-consciousness." That is, in studies where the critical comparison is between different states of consciousness, "state" may refer to the distinction between being awake, dreaming, or being in a dreamless sleep, or, alternatively, being responsive versus being in a coma, in a vegetative or minimally conscious state, or a locked-in

Liad Mudrik and Aaron Schurger, *How can we determine whether or not an agent is conscious of a bit of information relevant to an action?* In: *Free Will*. Edited by: Uri Maoz and Walter Sinnott-Armstrong, Oxford University Press. © Oxford University Press 2022. DOI: 10.1093/oso/9780197572153.003.0024

state. For such studies, *determining if someone is conscious* is done via some behavioral and physiological measures (e.g., the Glasgow scale, taken from the medical domain, where the level of responses to different types of stimulations are taken as measures for the patient's awareness level).

Conversely, *being conscious of a bit of information* is the focus of "content-consciousness" studies, where the information of which subjects are conscious is typically sensory. In such studies, participants are typically wide awake and responsive, and the manipulation pertains to a specific stimulus that is rendered invisible using different experimental techniques (e.g., masking, continuous flash suppression, crowding, and inattentional blindness, to name a few). Then one can determine whether the participant was conscious of that stimulus or not. But how can we do so? Two types of measures are routinely used. The first, *subjective measure*, is pretty straightforward: we ask the participant to rate the visibility of the stimulus (i.e., to report what they perceived). The second, *objective measure*, was developed as researchers realized that subjective report—despite being the most direct way to gain knowledge about the content of others' consciousness—might not always be accurate. Participants who have a strict criterion for being conscious (i.e., only say they are conscious of something when they clearly perceive it and report all cases of seeing something vaguely as unconscious) may say that they were not conscious of a stimulus even though they actually did see something. To validate their subjective reports, they are typically presented with a forced-choice judgment about that stimulus; if they are both (a) unable to consistently make that judgment correctly and (b) report not seeing the stimulus, researchers typically feel safe to infer that participants are unaware of the stimulus. Conversely, if only one of these conditions is met, but not both, they will *infer that participants may have been conscious of the stimulus (or of that bit of information).*[1]

Notably, this assumes that being conscious of information means having some access to that information (enough to enable correct judgments about it or to report it)—an assumption that has been

---

[1] See Chapter 12 by Block on consciousness causing physical action.

challenged by some philosophers and neuroscientists. Interestingly, recent studies have looked for ways to determine if participants are aware of specific information without relying on their report ("no-report paradigms"), using mostly physiological measures (e.g., an objective correlate of the awareness of that specific information. This could be a neural correlate, a pupillary response, or some other measure that typically correlates with subjects' conscious perception). This is taken as a means to pinpoint processes that are related to conscious processing per se, while excluding confounding processes related to the act of reporting (i.e., the neural correlates of, say, pressing a button could confound the neural correlate of the awareness of some specific information). Notably, however, these no-report measures are always ultimately proxies of participants' reports, as they were validated in previous experiments, where the measures were shown to be well correlated with explicit reports. That is, any physiological measure of consciousness was found in an experiment where researchers used a report paradigm; thus, by definition it correlates not only with conscious processing but also with the report and the processes underlying the report. And so the validity of no-report measures rests on their having been correlated with report in prior experiments. Thus, though these studies offer new and highly exciting possibilities for studying consciousness, they do not fully circumvent the problem of report.

Now we can go back to the original question: *How can we determine if an agent is conscious of a bit of information relevant to action?* Assuming that we know how to measure consciousness of a bit of information (as we discussed), we are left with the further question of what makes a bit of information relevant to action. If the subject is engaged in a task involving some kind of response instantiated by bodily movement, then arguably any sensory information relevant to the task is also relevant to action (the response). This is relatively unhelpful because it means that essentially any bit of information ends up being relevant to action. A better question might therefore be: When is a bit of sensory information *not* relevant to action?

Action-irrelevant information is a common occurrence in everyday life. When waving to a friend in a crowd, the shirt color of the person beside you may not be relevant to your action. But this is less common in experimental contexts, it turns out. First, experiments tend to use stimuli that are relevant to the task (except of course for experiments specifically focused on the effects of task-irrelevant information). And even when not all features of the stimuli are relevant, the irrelevant features tend to remain constant throughout the experiment. Second, just about any task in any experiment involves making a response in the form of motor output. So, at least indirectly, any bit of task-relevant information can be linked to an action in some way or another.

What we need in fact is a bit of information that sometimes *is* relevant to action and sometimes is *not*. This could potentially be addressed using a combination of a no-report paradigm and a standard subjective-report paradigm. The idea would be to image brain activity associated with the same information content both when that content is linked to a report and when it is not (notably, report is just an example of a task involving action; other tasks might be used as well). Then, for example using machine learning, we could try to learn the differential characteristics between neural activity carrying that information when it is relevant to action and when it is not (we would need to be careful to factor out activity that is purely related to producing the motor response. This can be done by randomizing the response mapping on each trial, so that subjects execute different motor responses to report having seen the stimulus and revealing that mapping only at the go signal, when subjects are asked to report. Then activity before the go signal would be abstracted from the motor response). Once we can identify the defining features of "action relevance" in neural information, we would be in a position to ask whether a given instance of such information is conscious (as discussed earlier). Prior research suggests that neural information is relatively more stable[2] (a property that can be measured) when it is conscious compared to when it is not. Using this or other agreed-upon "neural correlates of consciousness," we might then be in a position to ask whether a given bit of action-relevant information is conscious.

At present we still do not have an answer to the posed question; it is an open, active research question. We outlined an approach that might lead to an answer. Whether or not such an approach will work remains

---

[2] See Chapter 29 by Liljenström on stochastic neural processes.

to be determined empirically. But we hope to have laid out the necessary elements for making progress in that direction.[3]

Follow-Up Questions

Walter Sinnott-Armstrong

Philosophers often distinguish phenomenal consciousness from access consciousness, as in the chapters by Block and Bayne on consciousness in this volume.[4] Do the manipulations that you discuss—masking, continuous flash suppression, crowding, and inattentional blindness— manipulate phenomenal consciousness or access consciousness or both? Do the subjective and objective measures that you discuss measure phenomenal or access consciousness? Or does this distinction not make sense in the context of neuroscience?

Amber Hopkins

In Tim Bayne's answer to the question "How is consciousness related to freedom of action or will?,"[5] he suggests that immediate knowledge, or being conscious, of our intentions/decision is required for freedom (of action and the will). Would you consider decisions or intentions as bits of information relevant to action of which we can be conscious?

Jake Gavenas

There have been efforts to make subjective reports more objective (e.g., confidence ratings such as the meta-d' measure by Maniscalco & Lau,

---

[3] The authors of this chapter recommend Haggard, Clark, and Kalogeras (2002); Kim and Blake (2005); Reingold and Merikle (1988); Seth, Dienes, Cleeremans, Overgaard, and Pessoa (2008) for further reading.

[4] See Chapter 12 by Block on consciousness causing action and Chapter 13 by Bayne on consciousness and freedom of action or will.

[5] See Chapter 13 by Bayne on consciousness and freedom of action or will.

2012). Could such measures help consciousness research bridge the gap between subjective report and objective fact, or will such a gap exist until we have a solid understanding of the neurophysiological basis of consciousness?

Dehua Liang and Jye lyn Bold

What do you think about the idea of using machine learning/consciousness metrics (the level of neural integration, for example) to decide in real time whether a person is conscious or classify the different kinds of consciousness? And would such machine learning/consciousness metrics have to be individually tailored?

Replies to Follow-Up Questions

Liad Mudrik and Aaron Schurger

The questions we received are different, yet a joint theme to all of them concerns how we define and measure consciousness. Thus, we will first address this more general issue, while relating to specific points raised by each question. Since the seminal claim for unconscious perception by Peirce and Jastrow (1884), researchers have been struggling with finding good enough ways to measure consciousness and distill it from unconscious processes. Notably, in response to Gavenas, the goal of this endeavor was not to make subjective measures more objective but to make them more sensitive and accurate; the hallmark of subjective reports is that we cannot test them against a ground truth, as one does with objective measures. However, the only ground truth in subjective measures lies in what the subject actually experienced, and the only way to know what she experienced is to ask her. This also pertains to Sinnott-Armstrong's question about Block's distinction between access (A) and phenomenal (P) consciousness; since as experimenters we rely on subjective reports, we probe A-consciousness by definition.

Notably, there are means to infer what the subject experiences even without asking, using neurophysiological markers of conscious experience (which serve as the basis for "no-report" paradigms). But this would not make things more *objective*, as Gavenas suggested; it would simply be a different way of measuring the same thing (while importantly removing confounds related to the production of subjective reports). The reason for this is that these neurophysiological measures draw their validity from subjective reports to begin with, as they are considered markers of experience solely due to their correlation with reports in prior experiments. No matter how we turn it, studying consciousness is studying the subjective, not turning the subjective into the objective. But we can try to unravel the connections between the subjective and the objective, we can measure what is going on in my brain while I taste chocolate and examine the relations between that and my report of my experience, which *is* an objective fact—the report is observable and measurable irrespective of one's perspective.

This also applies to Liang and Bold's question about machine learning; this is a great idea and safe to say that it's been on people's minds ever since the resurgence of biofeedback in neuroscience research, under the guise of so-called brain-computer interfaces (BCIs). Mostly, BCIs have been used to try to read out the user's intentions and perform some action in response, like outputting text to a computer screen. But more recently researchers have devised metrics that, when applied to brain data and combined with machine learning, can be used to discern the level of consciousness of brain-injured patients (King et al., 2013; Schurger, Sarigiannidis, Naccache, Sitt, & Dehaene, 2015; Sitt et al., 2014). The holy grail of this kind of research would be to be able to read out in real time whether or not a human subject was conscious of a specific stimulus on a single trial and use that capability to deliver consciousness-dependent feedback in real time. We're not there yet; but even if we were, this would not solve the problem of relying on subjective reports, since, as we explained for neurophysiological markers, the BCIs pick up on neural differences correlated with report vs. no report, as this is the data on which they were trained.

Thus, going back to Sinnott-Armstrong's question, any time we do an experiment we are studying A-consciousness. But we could ask whether we are also probing P-consciousness. Are we? Well, one problem with this question is that it begs the further question of how we would ever know if we were. As Block himself acknowledges, in the vast majority of cases, A-consciousness and P-consciousness go hand in hand. When we have access to an experience, we also enjoy (or suffer from) its qualitative character. And when we report not perceiving a stimulus, we don't have any qualitative experience of it. The existence of A without P, and perhaps more so P without A, remains hotly debated (and the dissociation itself is far from being consensual and has evoked highly interesting discussions and empirical work trying to support or challenge it). Even supporters of the distinction would agree that such cases are very rare, and there is no reason to assume that they take place during the experiments we described earlier. There, either you experience a stimulus, and therefore can report on it, or you don't, and therefore there is nothing to report. Accordingly, our view is that though subjective measures indeed probe A-consciousness, in the paradigms we described accessibility is accompanied by P-consciousness. Even if one accepts the dissociation—and not all neuroscientists and philosophers do—it would be less relevant to the empirical study of consciousness, where the contrast is between the joint occurrence of A- and P-consciousness and the absence of both of them.

Finally, responding to Amber Hopkins, we note that whether or not consciousness of our intentions must precede an action in order for that action to be counted as an act of free will is a purely philosophical question.[6] Why? Because it entirely boils down to how we choose to define free will.[7] If we define an act of free will as the outcome of a conscious decision, then yes, consciousness will (by definition) be required. Most people would agree that conscious free will is the only kind of free will worth having, although there may be subtleties involved that go beyond the scope of this reply. The question posed here amounts to asking how we should define the word "intention" or "decision," and that is also a philosophical question.[8] We can define it however we like, but only once we have agreed on a working definition can we go about neuroscientifically looking for signs of such a thing in

[6] See Yaffe's replies in Chapter 1 on intention, Hieronymi's replies in Chapter 2 on the will, Chapter 7 by Hall & Vierkant on free will in degrees, Chapter 8 by Mele on free will, and Chapter 13 by Bayne on consciousness and freedom of action or will.

[7] See Chapter 5 by O'Connor on free will, Chapter 7 by Hall & Vierkant on free will in degrees, and Chapter 8 by Mele on determining if we have free will.

[8] See Chapter 1 by Yaffe on intention.

neural activity. One might consider an intention to be reflected in the sum total of neural activity antecedent to and causally efficacious for a specific action, but even that definition is fraught with problems. How to define "intentions" in neuroscientific terms remains a challenging problem[9] and one on which neuroscientists would do well to collaborate with philosophers. But in answer to your question, yes, we would consider decisions or intentions to be bits of information relevant to action of which we can in theory be conscious—but that remains an opinion.

[9] See Chapter 21 by Parés-Pujolràs & Haggard on intentional actions.

25

Which neural mechanisms could enable conscious control of action?

Jake Gavenas, Mark Hallett, and Uri Maoz

A lingering question in the neuroscience of volition pertains to how conscious experience might cause actions.[1] Everyday experience suggests that conscious thoughts and decisions lead to actions. However, is the subjective experience of these thoughts and decisions only epiphenomenal—like the sports commentator who can narrate the game without affecting it—or does conscious experience actually play a causal role in the process? Some neuroscientific findings[2] have often been interpreted as providing evidence against a role for conscious thoughts and decisions in the causal chain leading to action. And as more and more is understood about how unconscious brain mechanisms contribute to behavior, less and less space might be left for consciousness. Here we explore the perspective that consciousness[3] does have a causal role in bringing about and controlling proximal and/or distal actions. Thus, we try to explain what neural mechanisms could underlie a causal relation between the contents of consciousness (especially conscious willing) and the action to which they lead.

First, we consider *whether* consciousness contributes to action. On the one hand, it seems almost obvious that conscious experiences

---

[1] See Chapter 12 by Block on consciousness causing physical action.
[2] This refers mainly to the Libet experiments that are discussed in more detail in Chapter 19 by Bold et al. and Chapter 28 by Lee et al.
[3] The word "consciousness" refers to at least two different concepts: first, the state of being conscious (as opposed to unconscious); second, one's ongoing subjective experience. The state of being conscious is certainly related to action in some way; we usually stay more still when sleeping or comatose than when awake, for example. But the mind-body problem in this context refers to how the *contents* of consciousness (also called qualia or subjective experience) can cause actions.

Jake Gavenas, Mark Hallett, and Uri Maoz, *Which neural mechanisms could enable conscious control of action?* In: *Free Will*. Edited by: Uri Maoz and Walter Sinnott-Armstrong, Oxford University Press. © Oxford University Press 2022. DOI: 10.1093/oso/9780197572153.003.0025

of willing or pain cause bodily movements, and that conscious experiences of happiness or sadness affect our behavior. On the other, the mind-body problem makes the question thorny. How can subjective experiences affect the objective world? Nevertheless, subjective consciousness is instantiated in the objective brain, likely as particular patterns of brain activity.[4] Furthermore, present activity affects future activity, potentially including activity in motor cortex. Thus, it is certainly possible that consciousness contributes to action. Finally, given evolutionary constraints, consciousness would probably not evolve if nonconscious organisms were behaviorally equal to conscious organisms. Therefore, it is not only possible but likely that consciousness contributes to action in some way.

Next, we consider *how* consciousness contributes to action. One way to begin is to trace the causal chain for action backward from bodily movement.[5] Muscle contractions cause overt, external actions. For voluntary action, the spinal cord and brain stem control muscles, and those are controlled by higher-level[6] brain regions. However, to go further we must make some assumptions about how consciousness is instantiated in the brain. Hence, we now turn to existing theories and frameworks for subjective experience and consciousness. Although most theories of consciousness focus on perception, several address action specifically. Two broad dichotomies can help categorize these frameworks: first, whether consciousness plays an immediate or distal role in action; second, whether consciousness itself generates actions or modulates actions generated by unconscious processes. Classical thinking would say that consciousness is immediate and generative: our conscious action plan initiates and drives motor execution. Libet's veto mechanism is a classic example of immediate modulatory control, proposing that consciousness can "kick in" and inhibit actions that are generated by unconscious activity. Examples and the location of some theories on this chart are given in Table 25.1.

---

[4] We make a physicalist assumption, like most of neuroscience: consciousness is comprised of brain activity.

[5] See Chapter 16 by Haggard & Parés-Pujolràs on the neural processes that produce actions and Chapter 20 by Hallett on levels and control in the brain.

[6] For more on such hierarchies in the brain, see Chapter 20 by Hallett.

Table 25.1 Two dimensions for differing views on conscious control of action: (1) Consciousness could influence immediate actions (proximal) or actions in the future, e.g., by planning (distal); (2) consciousness could act as a "source" for actions, generating or helping to generate them, or could act as a filter, modulating actions "offered up" by unconscious processes.

	Generative	Modulatory
Proximal	Classical views, active inference	Veto mechanism/response inhibition, passive frame theory
Distal	Planning	Reinforcement learning, distributed adaptive control

An overview of action-based theories of consciousness, as well as a more in-depth analysis of consciousness and action, is given elsewhere (Seth et al., 2016). In brief, four existing action-based theories of conscious stand out: Bayesian brain/active inference, sensorimotor contingency theory, distributed adaptive control, and enactive autonomy. Notably, active inference (an extension of Bayesian brain/predictive coding) argues that perception and action utilize the same underlying computations: minimizing prediction error of incoming sensory information (Adams, Shipp, & Friston, 2013). In perception, incoming sensory information is compared to a top-down prediction, and the error is used to update an internal model. Comparatively, in action, top-down predictions are made, and muscles contract to minimize the error in those predictions. And although active inference is not explicitly a theory of consciousness, it is thought that the top-down predictions may be related to subjective experience (more so than bottom-up information processing). Therefore, the computations underlying prediction and minimizing prediction error may form a bridge between perceptual consciousness and volitional action generation.

Distributed adaptive control also proposes a multilevel, hierarchical control scheme, where the state of the world, the self, and potential actions are modeled and guide behavior. However, under this theory, consciousness plays the role of an interpreter, sorting ongoing events in order to guide future action, and thus this theory might predict that consciousness influences action through mechanisms related to reinforcement learning. Both sensorimotor contingency theory and enactive autonomy argue that conscious experience is related to action-based engagement with the environment but are more focused on psychological rather than neural mechanisms.

Another relatively new framework put forward by Morsella, Godwin, Jantz, Krieger, and Gazzaley (2016) is passive frame theory. They argue that consciousness helps resolve motor indecision when different neural systems are attempting to output conflicting actions (for instance, when one is holding one's breath underwater). Passive frame theory proposes that consciousness is an information bottleneck that integrates activity from across the brain. Information from different modalities such as memory or vision would interact in the stream of consciousness, resulting in the optimal action being expressed by the motor system. The theory thus posits that consciousness modulates immediate action production.

Although not a theory of consciousness per se, optimal feedback control (Scott, 2004) offers an explanation for motor control that fits well with generative frameworks for conscious control. In broad strokes, potentially conscious goals are generated in prefrontal areas, with input from limbic system and basal ganglia, among other structures. Motor cortex then translates neural representations of goals into motor commands. Copies of these commands are maintained to predict action outcomes and then compared to actual outcomes in parietal networks (Hallett, 2016; Shepherd, 2015). Differences between predicted and real outcomes are then used to guide further motor commands in a feedback loop. This might give rise to the experience of having caused an action— the sense of agency. In this model, consciousness may play a generative role in deciding which goals to pursue, and thereby which actions should be taken. However, under optimal feedback control, consciousness may also play a modulatory role, determining which sorts of feedback are prioritized to adjust actions (Shepherd, 2015).

Several studies have also investigated how consciousness contributes to action outside of the context of any particular theory, focusing instead on how consciousness interacts with particular neural mechanisms. For instance, motor cortex excitability increases with conscious intention (Zschorlich & Köhling, 2013), which could be a mechanism for immediate control of action, generative or modulatory. Furthermore, perceptual masking of a reward has been shown to lower, but not eliminate, humans' ability to learn from actions (Correa et al., 2018). Flipping this around, consciousness may assist in learning from prior actions, which may be a form of distal and modulatory control of action. Finally, perceptually masking stop signals lengthened participants reaction times (van Gaal et al., 2009), suggesting that immediate inhibitory control mechanisms do not require consciousness. The same study also showed that conscious errors led to slower responses on the following trial while unconscious errors did not, implying that distal inhibitory control mechanisms may require consciousness. Findings such as these guide action-oriented theories of consciousness by pointing towards and away from particular neural mechanisms. However, it is unclear whether such studies investigate subjective experience itself or its correlates, which is why further theory-driven research is needed to ascertain the relation between consciousness and action.

We focused here on what neural mechanisms might allow conscious, subjective experiences (or their neural implementations) to control action. Further efforts are needed (1) to synthesize previous and new findings with extant frameworks, (2) to expand current frameworks or to develop new ones to account for these findings, and (3) to integrate action-based with perceptual frameworks. There is general agreement that humans have conscious, subjective experiences.[7] But further advances in the understanding of conscious control of movement are required to explain how those experiences lead to action—a key question in the neuroscience of volition.

Follow-Up Questions

Amber Hopkins

You touch on the idea that consciousness likely serves a functional purpose because otherwise it would not have evolved. However, could it be that the kind of brain we have was selected for rather than consciousness? The brain

---

[7] Though *Meditations* by René Descartes (1641/1984) provides a skeptical counterargument.

is a complex, specialized, and interconnected system that supports sensory integration, cohesive perceptions, complex behavior, and planning, among many other processes. Organisms with a brain like this would surely behave differently, possibly more adaptively and successfully, than those without. Given that the brain is such a complex and interconnected system, consciousness could be an emergent property or byproduct that was not specifically "selected for." So could it be that the kind of brain we have was "selected for" and that consciousness is an emergent property of this kind of brain with no "selected for" functional purpose?

Timothy O'Connor and Adina Roskies

Are there any (action-driving) functions that are either unique to or enhanced by consciousness? For example, responding to reasons, logical inference, or metacognition, where relevant to action.

Walter Sinnott-Armstrong

In his chapter, Ned Block proposes a model of how the conscious aspect of a mental event can cause a physical action.[8] Are his views compatible with the theories that you discuss, especially distributed adaptive control and passive frame theory? More generally, how are philosophical theories of how causation by consciousness is possible related to neuroscientific accounts of how such mind-body causation works?

Replies to Follow-Up Questions

Jake Gavenas, Mark Hallett, and Uri Maoz

First, we will respond to the question and argument offered by Amber Hopkins regarding the evolutionary argument for consciousness

---

[8] See Chapter 12 by Block on consciousness causing action.

having a biological function. Let us start by stating that our response is highly speculative. Now, restating the question, perhaps subjective experience—the phenomenological aspect of consciousness—merely became a property of the brain as an evolutionary byproduct of a complex nervous system rather than being selected for directly. This is certainly possible: evolutionary pressures bias the human genome toward genes that express adaptive traits. Nevertheless, the human genome does not specify every connection in the brain (see Zador, 2019); rather, evolutionarily selected genes provide the scaffolding for the brain and consciousness to develop.

However, evolutionary perspectives may still favor consciousness having a biological function. We continue to assume physicalism is true, which boils down to the assumption that consciousness is physically instantiated in the brain and is equivalent to or realized by some patterns of neural activity (similar to the relationship between computer hardware and software; Verschure, 2016). Consciousness is influenced by incoming sensory and proprioceptive information (because wakeful, subjective experience generally tracks objective reality—though note exceptions such as optical illusions). So conscious neural activity receives input from sensory regions. Furthermore, because subjective experience is intricate, this activity is likely non-trivial: a large swath of neurons is probably required to encode conscious content. An architecture where complex input is received and processed yet no function is served is unlikely to have evolved in the first place, *especially* given that the neural connectome is not genetically specified. Moreover, if such conscious neural activity did not serve any biological function, it would actively *waste* energy and thus be evolutionarily *disadvantageous*. So, from an evolutionary perspective, we believe it is likely that consciousness serves a functional purpose.

Next, we turn to the similar question offered by Timothy O'Connor and Adina Roskies: If consciousness does have a function, what is it? Whether any function in particular is unique to consciousness remains an open question. Some argue that everything that happens above the consciousness threshold can also happen below it (see Hassin, 2013), albeit perhaps at decreased efficiency or effectiveness. Others claim that strictly unconscious processing is, in fact, quite limited (Newell

& Shanks, 2014). Between these extremes lie various perspectives. The answer is not, at least yet, known, but there has been a good deal of speculation.

An abundance of research demonstrates that myriad processes can take place unconsciously (although the extent to which these processes are unconsciously processed in day-to-day life is unclear). Certain perceptual processes can occur without awareness, as evidenced, for example, by "unconscious vision" in blindsight (Ajina & Bridge, 2017). Semantic information can be processed unconsciously (Mudrik, Breska, Lamy, & Deouell, 2011), as can emotional information (Liu et al., 2016). Even attention, which may be critical for some information to reach consciousness, can be directed toward stimuli without awareness of those stimuli (Van Boxtel, Tsuchiya, & Koch, 2010).

Integration lies at the nexus of mainstream theories of consciousness, including global neuronal workspace theory and integrated information theory. Integration combines information from various sources into a coherent whole, such as merging visual and auditory stimuli to understand the visual source of a sound (multisensory integration). Even though some forms of integration can occur outside conscious awareness (Faivre, Mudrik, Schwartz, & Koch, 2014), consciousness may be required for or enhance more complex forms of integration (Mudrik, Faivre, & Koch, 2014). For instance, in a phenomenon known as the McGurk effect, visual information from mouth movements can influence auditory experience (e.g., hearing "fa" instead of "ba" for the same sound if you see a mouth clearly articulating "fa"). Suppressing consciousness of mouth movements using continuous flash suppression eliminates the McGurk effect (Palmer & Ramsey, 2012), suggesting that consciousness enhances multisensory integration.

Of functions relevant to volition, responding to reasons is therefore a strong candidate for enhancement by consciousness. Reasons-responsiveness likely relies on the integration of memory, evaluation, planning, and potentially motor execution processes. Furthermore, passive frame theory proposes that consciousness helps resolve conflict when multiple neural systems simultaneously attempt motor output. Thus, consciousness may play a particularly important role when one is responsive to multiple, especially conflicting reasons.

Finally, we turn to Walter Sinnott-Armstrong's question about how several theories we introduce interface with a model of conscious causation offered by Ned Block.[9] Block argues, first, that unconscious processes almost always precede conscious ones, yet conscious decisions may still play a causal role in behavior based on a counterfactual analysis. In particular, if the conscious aspect of a psychological or brain state were removed (or changed), and one's action were different as a result, we could consider the conscious aspect as causative. Block likens this to the iceberg that sank the *Titanic*: Did the underwater part of the iceberg cause the ship to sink? Without it, the ice that pierced the ship's hull might have failed to do so, and thus we could consider the underwater portion of the iceberg to have causally contributed to the *Titanic*'s sinking.

Now, we explore two action-oriented theories of consciousness with Block's counterfactual model. First, distributed adaptive control proposes that consciousness serves a more distal function, where the causal contribution to one action in particular is less clear. Distributed adaptive control argues that consciousness optimizes the *future* control of action by representing and evaluating current performance (Verschure, 2016). So if a specific conscious content optimizes (i.e., affects) a future action, we would assume that lacking that content, the future action would be different (or perhaps change the probability distribution of actions). For instance, if you are practicing piano and you consciously recognize a misplayed note, you might adjust your playing the next time around—whereas, if you are not conscious of the misplayed note, your future actions would not be adjusted.

Next, passive frame theory proposes that consciousness is an integrative bottleneck that allows for the adaptive resolution of motor conflict. Conflict may stem from unconscious recruitment of motor regions. Notably, unconscious recruitment of motor regions may begin before awareness of conflict. If the conflict were unconscious, the conflict may be resolved by whichever recruitment is immediately stronger, precluding a more nuanced response. For instance, when ordering lunch, you may have a strong urge to say "burger." You may

---

[9] See Chapter 12 by Block on consciousness causing action.

also have a weaker urge to say "salad." If conflict between these motor urges remains below the threshold of consciousness, the stronger desire to say "burger" might proceed akin to a habit. But if motor conflict surmounts the threshold, consciousness of both urges would emerge, thus prompting you to reconsider your option and, potentially, alter what you say.

In terms of conscious causation in general, physicalism remains the prevailing assumption in neuroscience, whereas philosophers embrace a wider variety of perspectives (a natural consequence of arguing from first principles and phenomenal experience. Even though few scientists explicitly endorse dualism, it is implicit in several explanations and theories (see Mudrik & Maoz, 2014 and Blackmore's commentary in Morsella et al., 2016). From the physicalist perspective, however, the problems of conscious causation reduce to the mind-body problem: How is consciousness instantiated in the brain, and how should we theoretically reconcile subjective, first-person experience with objective, third-person information about brain activity?[10]

---

[10] The authors of this chapter thank Liad Mudrik for useful comments on an earlier version of this manuscript.

26

How does the absence of a consensus about the neural basis of consciousness and volition affect theorizing about conscious volition?

Amber Hopkins, Liad Mudrik, and Uri Maoz

Many think that science is an endeavor to seek truth and acquire knowledge; fewer contend that science primarily aims to achieve consensus. Indeed, consensus is neither necessary for scientific progress nor does it represent the veracity of contemporary scientific knowledge and practices.[1] Robust empirical data, precise language to describe data, and coherent frameworks within which to interpret data are key to scientific progress.[2] However, collective agreement and shared (explicit and implicit) assumptions constitute the foundation from which scientists ask questions and devise paradigms to answer them. Thus, similarly to how averaging over many surveys tends to result in better predictions than a typical individual survey (Silver, 2012), consensual scientific opinions have a bigger claim on the truth than an arbitrary, dissenting opinion.

Let's take the mechanisms that give rise to conscious vision as a case in point. There is relative consensus that photoreceptors transduce light into electrical signals that travel by way of the lateral geniculate nucleus to the primary visual cortex and then undergo additional processing. At some point in this process, and due to some neural-network

---

[1] If anything, the history of science is riddled with prevailing, albeit false, consensuses that hindered progress: geocentrism, humorism and biles, phlogiston, and many more.
[2] See Kreiman's replies in Chapter 17 on the neural basis of the will.

Amber Hopkins, Liad Mudrik, and Uri Maoz, *How does the absence of a consensus about the neural basis of consciousness and volition affect theorizing about conscious volition?* In: *Free Will.* Edited by: Uri Maoz and Walter Sinnott-Armstrong, Oxford University Press. © Oxford University Press 2022. DOI: 10.1093/oso/9780197572153.003.0026

activation, these signals are refined and combined with other information in such a way that conscious visual experience arises. The imperative, or fundamentally necessary, generating mechanisms of this higher-level phenomenon are neither well-understood nor agreed upon, and thus hypotheses diverge. Yet the relative consensus on the basic dynamics of the visual system serves as a basis, allowing scientists to debate, investigate, and hypothesize about higher-level phenomena. Thus, some consensus is indeed constructive for making progress, but investigations can continue even in the absence of full consensus.

In a similar manner, neuroscientists of volition start from a consensus on some neural mechanisms involved in sensation, perception, and action production, but then diverge regarding more abstract and complex processes or phenomenological states associated with volition. We assume some low-level phenomena, for example that the primary motor cortex is somatotopically organized, that descending and ascending pathways along the spinal cord allow for information flow between the brain and the body, and that motoneuron activation leads to muscle contraction.[3] However, we then craft unique methodologies and cognitive or computational models to investigate corresponding higher-level phenomena related to volition, such as deciding, intending, initiating, controlling, awareness, consciousness, and the causal relations between them.

Presently, we lack consensus about the neural underpinnings of consciousness. The mechanisms that give rise to consciousness are not well understood and certainly not agreed upon. A non-exhaustive list of neurobiological frameworks includes first-order theories, higher-order ones, recurrent processing theory, thalamocortical loops theory, attention schema theory, global neuronal-workspace theory, and integrated information theory. Each such framework posits its own set of essential qualities of consciousness, the necessary or sufficient neuronal properties that generate it, and lays foundations from which we can derive framework-specific hypotheses. Frameworks, along with philosophical analyses, innovative methods, models, and data, illuminate the relation between the brain and consciousness.

---

[3] See Chapter 20 by Hallett on levels and control in the brain.

The essential features of volition, what is required for volition, its neural basis, and its role in action are also not wholly agreed upon. Volition can be understood as a state or process that leads to, or as the capacity for, endogenous and goal-directed action (Haggard, 2019). Intuition may suggest that we should be able to reverse-engineer the voluntary motor pathway back from action with the goal of identifying the initial action-generating command in the brain, or the neural basis of the volitional state, that we surmise first causes the cascade of events leading to movement. However, it is unclear what we should be looking for in the brain. The neuroscience of volition has emphasized one voluntary-movement-preceding signal: the Bereitschaftspotential (BP), or readiness potential (RP). But the nature of the RP remains controversial (Fried, Haggard, He, & Schurger, 2017). Even more, processes involved in volition, such as deciding, intending, initiating, and controlling, are themselves abstract. In particular, deciding and intending[4] are diverse and dynamic processes, making their neural instantiations and role in generating or modulating action elusive (Schurger & Uithol, 2015). However, neuroscientists make continued progress despite lacking consensus. Developing precise language to distinguish the relevant concepts, enhancing our research methods, as well as collecting and modeling data, allow us to better discern the neural underpinnings of volition and its contribution to decisions and action.

As there is no consensus regarding both consciousness and volition, it might come as no surprise that the combined phenomenon, conscious volition, poses a special challenge. Volition and consciousness seem related. Volitional, or voluntary, acts seem to have subjective and self-referential qualities—they feel "up to us."[5] Some claim that to act volitionally we must be aware that we are acting and feel that our action is the result of our conscious decision, intention, initiation, and control. However, whether or not consciousness is necessary for volition, it remains unclear when, if at all, consciousness enters the causal chain leading to action.[6]

---

[4] See Chapter 1 by Yaffe on intention and Chapter 21 by Parés-Pujolràs & Haggard on intentional actions.

[5] See Chapter 21 by Parés-Pujolràs & Haggard on intentional action.

[6] See Chapter 25 by Gavenas et al. on conscious control of action.

Crucially, lacking consensus on consciousness and volition does *not* prevent us from making scientific progress, theorizing about conscious volition, or acquiring knowledge.[7] The search for neural correlates of high-level phenomena—such as consciousness or conscious intentions—is in a fairly exploratory phase. Dynamics, mechanics, and spatiotemporal neural correlates for these more complex phenomena have been proposed, opening such ideas up for dissection, rebuttal, or support. Progress can be made irrespective of framework affiliation or metaphysical position. Consensus is not necessary to explore these concepts neurophysiologically or to search for their neural correlates. If anything, the reverse is true: these investigations may lead to consensus in the future, with more empirical data that supports or challenges one of the theories or gives rise to new ones.

Appealing to Kuhn's (1962) idea of a paradigm shift, sometimes breaking consensus can lead to greater progress. Examining the history of science, Kuhn theorized that scientific development occurs in phases. During "normal science," the scientific community adheres to the dominant paradigm—assumptions, methodologies, and a set of "puzzles" to be solved in "acceptable ways." Progress on solving these puzzles leaves scientists confident in the dominant paradigm. But the accumulation of enough puzzles that cannot be resolved within the dominant paradigm can eventually result in a communal shift to a new paradigm and a new phase of normal science. And the cycle continues.

It should also be noted that consensus is lacking not only for the neural underpinnings of consciousness and volition but also for high-level concepts like attention, emotion, memory, language, and more. If anything, it is debate and mutual critique among different theories of these concepts that drive discovery in these fields.

In sum, consciousness and volition are very complex concepts. They have eluded consensus over millennia of philosophical investigation and now perplex neuroscientists, who relatively recently joined this effort. However, both philosophy and science have made progress during this time, without consensus. And there is no reason to think that the same will not happen in this case.

---

[7] See Chapter 9 by Roskies on neuroscientific evidence about free will.

Follow-Up Questions

Walter Sinnott-Armstrong

Philosophy as a field is often criticized for failing to reach consensus. You tell us that science is in the same boat but can still make progress. Do you see any difference between philosophy and science that would make the failure of consensus a bigger or smaller problem in one field than in the other? If a lack of consensus does not preclude progress in science, as you argue, is the same true in philosophy?

Mark Hallett

Would you be willing to say that the fact that volition (or at least the usual consequence of volition—movement) is measurable makes it possible to study even if we do not understand consciousness?

Dehua Liang

To what extent does philosophy have a role in helping neuroscientists reach consensus about the neural basis of consciousness and volition? In the last paragraph, you mention that science and philosophy have made progress together in the past. You also mention at the beginning of the chapter that science is providing robust empirical data and models to describe mechanisms. If neuroscientists are eventually able to reach a consensus about consciousness and volition with better data, what would be the role of philosophy in this process?

Deniz Arıtürk

Does a neuroscientific theory's success depend on the philosophical theory of consciousness it is judged against? Do different neuroscientists have commitments to different philosophical theories that affect their experimental framework and evaluation of alternative neuroscientific theories? What, if any, are the philosophical assumptions about consciousness that all neuroscientists rely on that not all philosophers agree on?

Replies to Follow-Up Questions

Amber Hopkins, Liad Mudrik, and Uri Maoz

This book is rooted in the fruitful relations between neuroscience and philosophy and aims to foster interdisciplinary collaboration and discussion. As such, we would like to emphasize, in response to Dehua Liang, that philosophy is an indispensable companion to neuroscience when investigating high-level concepts like consciousness and volition. It offers conceptual clarity, helps form relevant empirical questions, aids in the appropriate design of experiments to answer such questions, and, once results are obtained, helps with their interpretation. Philosophical clarity is crucial when investigating intricate phenomena like consciousness and volition empirically, as even the language used to discuss these terms can be fraught (Mudrik & Maoz, 2014). Needless to say, philosophy also stands on its own. It elucidates numerous issues outside the scope of scientific investigation (e.g., metaphysical questions).[8] Hence, philosophy does not only "provide an essential service" to science but also complements science where it is unable to provide answers.

Bearing in mind that philosophy and neuroscience are independent disciplines, and in response to Deniz Arıtürk, it is not surprising that these fields address the problem of consciousness differently. Neuroscientific theories of consciousness are mostly focused on pinpointing the neural mechanisms and brain dynamics that allow for conscious experience. So their content is the brain, neurons, neuronal activity, measurements or data, models, and predictions. Philosophical theories of consciousness mostly focus on the metaphysics, or existence, of consciousness, its essential features, how it should be defined,

---

[8] See, for example, Chapter 6 by O'Connor on determinism.

and critical debates with arguments and positions thereof. While there is some overlap, these fields largely focus on different problems and content.

There are accordingly different neurobiological and philosophical theories of consciousness, with no proper one-to-one correspondence between them (although some neuroscientific accounts are based on philosophical positions, such as higher-order thought theories, or on phenomenology, like integrated information theory). Neuroscientific theories (and models)[9] are typically judged not against philosophical theories but against empirical findings: if the latter do not confirm the theory's predictions, then the theory should either be revised or excluded altogether. Similarly, a philosophical theory's success is not determined by its compatibility with a neuroscientific theory but by how well it stands up to criticism and counter-argumentation.

As for the commitment of neuroscientists to philosophical stands, though we base this suggestion on no empirical findings, we think we can safely estimate that most contemporary neuroscientists are sympathetic toward physicalism and reductionism. Many neuroscientists assume that consciousness is the product of neural activity and is by itself physical. Additionally, many neuroscientists favor a reductionist, or methodologically reductionist, view, which is based on the belief that whole systems are best explained in terms of their constituting parts. In neuroscience, we explain the whole brain, a system, and its dynamics in terms of neurons, the brain's constituting parts, and their electrochemical processes.

In response to Walter Sinnott-Armstrong, some suggest that there has been more convergence on answers to questions about the world, or consensus, in the sciences than in philosophy (Chalmers, 2015). It is worth clarifying that data seems to be the catalyst for the greater tendency toward convergence on answers in science than in philosophy. Accordingly, it is often said that in science "data is king." What this means is that the results from empirical data that were properly collected (using agreed-upon methods that meet the standards) and then properly analyzed are considered the final arbiter of scientific

---

[9] See Chapter 30 by Kreiman et al. on computational models and free will.

disputes. This is not to say that scientific debates have not lingered, with each side brandishing what it thinks are decisive empirical data that promotes its theory. However, often one experiment or a set of experiments does end up providing decisive evidence in favor of one theory over others. And that experiment is viewed, at least in retrospect, as the watershed moment for the formation of a consensus. Famous examples include Eddington's measurements during the 1919 solar eclipse that demonstrated the superiority of Einstein's general relativity over Newton's or Hubble's confirmation that nebulae are galaxies and not objects inside the Milky Way. Current attempts related to neuroscience include the search for agreed-upon experiments that would be accepted as decisive among proponents of different scientific theories of consciousness. Lacking such decisive data, theories often multiply and speculation abounds. So, in science, data that would help drive consensus is actively sought and often viewed as progress.[10]

To the extent that neuroscientists are in a position to reflect on progress in philosophy, there seem to be instances of convergence and agreement in philosophy. However, rather than prioritizing convergence and agreement, philosophical arguments seem to result in "sophisticated disagreement" (Chalmers, 2015, p. 15). The critical engagement with – or criticism of—ideas and positions in philosophy does not result in the elimination of those ideas or positions, but instead provokes their enhancement and refinement. Philosophical progress might thus come about by better clarifying deep questions and concepts—like free will and consciousness. This need not, and often does not, lead to a consensus among philosophers. And such a consensus might not be sought. When it comes to cognitive neuroscience and the philosophy of mind, convergence on answers to "big questions—the problems of consciousness and intentionality, of mental causation and free will— [are] wide open" in both disciplines (Chalmers, 2015, p. 19).

Last, focusing on neuroscience, Mark Hallett suggests that the measurability of movement might facilitate the study of volition, even given our relatively poor understanding of consciousness. In response, we first state that we are not sure that volition is necessarily more

----

[10] Though it should be noted that there are those who do not see such clear progress in science (a famous example is Feyerabend, 1993, *Against Method*).

measurable, via movement, than consciousness. Distal decisions (e.g., deciding now to go to the party next week) do not result in any immediate contraction of muscles. Other decisions never result in contraction of muscles (e.g., a proximal decision to think now about a problem one is facing at work while driving home from work rather than allowing one's mind to wander). What is more, volition and movement can be dissociated (see, e.g., Desmurget et al., 2009). Paralyzed individuals and those suffering from locked-in syndrome seem able to consciously experience at least some of the following mental states related to volition: desiring, deciding, intending, and initiating a movement, without producing action. And, on the other hand, volition and consciousness are much harder to dissociate; we tend not to ascribe volition to unconscious individuals. And the subjective, internally generated, self-caused experience of acting voluntarily is sometimes interpreted as, at least part of, what makes an act volitional. To be clear, despite all of this, we hold that we certainly can study volition, as well as consciousness—though both are challenging endeavors, conceptually and empirically, and neither is necessarily easier to study than the other.

PART II, SECTION IV

QUESTIONS ABOUT
NEUROSCIENCE METHODS

27

How can we determine the precise timing of *brain* events related to action?

Mark Hallett and Aaron Schurger

Human neuroscience offers a wide array of techniques for measuring and recording brain activity. These include invasive techniques like single-unit and local-field-potential recordings, and non-invasive techniques like electroencephalography (EEG) and functional magnetic resonance imaging (fMRI). Recording the precise timing of brain events requires recording techniques that can keep pace with neuronal events that play out on the time scale of milliseconds. Brain events related to action, such as bursts of neural activity in primary motor cortex leading to a muscle contraction, operate on a time scale of 10s of milliseconds. Neural activity in the supplementary motor area (SMA) can precede movement by up to a full second and may be similarly extended in time. In both cases, however, if we want precision, then we need recording techniques that measure the timing of events by the millisecond.

fMRI is a very popular and powerful technique, allowing us to non-invasively measure brain activity with millimeter-level spatial resolution. But it is slow, typically recording one sample of blood oxygen consumption in the brain every 1.5 or 2 seconds. This is not at all fast enough to precisely capture neural events that lead to the initiation of action. The same is true of techniques like positron emission tomography and single-photon emission computed tomography. EEG measures brain activity on the millisecond level, but it measures that activity in the aggregate because only large-scale synchronous neural events have any chance of being picked up as changes in the electrical field

Mark Hallett and Aaron Schurger, *How can we determine the precise timing of the* brain *events related to action in humans?* In: *Free Will.* Edited by: Uri Maoz and Walter Sinnott-Armstrong, Oxford University Press. © Oxford University Press 2022. DOI: 10.1093/oso/9780197572153.003.0027

at the scalp. Its temporal resolution is high, but its spatial resolution is relatively low. It is also hard to determine the source of the neural activity from EEG because the electrical field is heavily distorted by the cerebral-spinal fluid surrounding the brain, by the skull bone, and so on. Among the non-invasive methods, perhaps the best compromise between spatial and temporal resolution is magnetoencephalography (MEG), which measures very weak magnetic fluctuations that are induced by electrical discharges in populations of neurons. MEG has the same temporal resolution as EEG but somewhat superior spatial resolution. Still, even MEG suffers from serious limitations. It is relatively insensitive to deep (sub-cortical) sources and can completely miss activity on the crests of gyri (because of the orientation of the electrical fields with respect to the magnetic sensors), though the latter can be ameliorated by concurrently recording MEG and EEG.

It is possible to measure neural activity directly using microelectrodes to record the spiking activity of single neurons or the local field potential (LFP) in a small brain region, but, for ethical reasons, this can be done in human subjects only when there is a specific medical reason for implanting the electrode(s), such as monitoring epileptic activity. Still, studies of human subjects with depth electrodes (an electrode placed into the brain) recording single-neuron activity have been carried out. In one such study (Fried, Mukamel, & Kreiman, 2011), the authors were able to directly record the brain activity leading up to self-paced actions performed in the context of Libet's (Libet, Gleason, Wright, & Pearl, 1983)[1] self-initiated movement task. In this study the authors were able to capture a buildup in neuronal activity, at the level of single neurons, that led up to the initiation of a self-paced movement (and the feeling of intending to move). Another invasive recording technique, electrocorticography (ECoG), captures aggregate activity directly on the surface of the cortex. While not as spatially precise as single-unit recordings, ECoG can cover a much greater spatial expanse, covering large areas of cortex and enabling us to capture widespread activity that might be functionally

---

[1] For further details about this experiment, see Chapter 19 by Bold et al. and Chapter 28 by Lee et al.

precise vis-à-vis the initiation of movement but not isolated to a single small spatial focus.

In addition to recording methods, it is possible to determine the timing of brain events with stimulation methods. Again, this can be done invasively or non-invasively. A popular non-invasive method is transcranial magnetic stimulation (TMS). For example, a single pulse of TMS delivered focally over the cortex can briefly inhibit neural activity in that region of cortex. This could cancel or delay a neural process utilizing that region of cortex that is occurring at that time. A TMS pulse over the motor cortex during a reaction-time task can delay the response; a TMS pulse over the SMA during a motor sequence can disrupt the sequence.

The ability to measure brain activity on a precise time scale is certainly important, but there are other considerations regarding the question posed here. Related questions are whether the brain activity related to action happens at a precise time and whether the relevant activity might be spread out in time (a process) rather than being a point in time (an event). In self-paced movements, the EEG shows a slow buildup in the 1.5 to 2 seconds prior to the movement, called the Bereitschaftspotential (BP, or readiness potential). Comparing the EEG to ECoG shows that the early BP arises from activity in the SMA and premotor cortex and the later BP adds activity from the primary motor cortex culminating in the burst of activity that drives the movement. The onset time of the BP is typically measured as the time at which the average signal deviates significantly from the baseline level of noise. Thus, while the time of the peak of the late BP can be estimated with precision, the same might not be true of the onset of the early BP, since it depends in part on the signal-to-noise ratio (i.e., the number of trials being averaged). Also, the onset time of the BP is highly variable across studies and between subjects. It can begin anywhere from 2 seconds before to 0.5 seconds before movement onset, depending on the study and individual subject. This means that, even if we record the BP with millisecond precision, determining its onset time might be problematic if it is spread out in time rather than being a punctate event.

In summary, determining the precise timing of the brain events related to action requires a recording method with millisecond precision

(EEG, MEG, ECoG, single unit, or LFP) and also hinges on whether the brain activity happens at a precise time or is extended in time. Several recording techniques offer millisecond time resolution and could, in principle, be used to this end. The most well-studied neural antecedent of self-initiated action, the BP, begins up to 1 second or more before action onset and builds up slowly. The precise onset time of this pre-movement buildup is highly variable between subjects and studies and is difficult to estimate. Future studies should focus on this very important question.[2]

Follow-Up Questions

Walter Sinnott-Armstrong

One reason why timing of brain events is important is that causes cannot come after their effects, so timing is supposed to help us sometimes get from correlations to causation or at least rule out causal hypotheses when we find that the hypothesized cause occurs later than the hypothesized effect. Do neuroscientists care about causation? Do they accept the philosophical assumption that causes come either earlier than or at the same time as their effects but never come after their effects? Can the methods of timing brain events that you discuss be used to rule out any otherwise plausible causal hypotheses? In particular, many philosophers claim causal roles for the will (or choices), as in Hieronymi's chapters,[3] intention, as in Yaffe's chapter,[4] and consciousness, as in Block's chapter[5] and Bayne's chapter.[6] Can the methods of timing brain events that you discuss be used to test any of these causal hypotheses? If so, how?

---

[2] The authors of this chapter recommend Fried et al. (2011); Gross (2019); Schneider, Houdayer, Bai, and Hallett (2013); Shibasaki and Hallett (2006) for further reading.

[3] See Chapter 2 by Hieronymi on the will and Chapter 3 by Hieronymi on voluntary action.

[4] See Chapter 1 by Yaffe on intention.

[5] See Chapter 12 by Block on consciousness causing action.

[6] See Chapter 13 by Bayne on consciousness and freedom of action or will.

Adina Roskies

How does Schurger, Sitt, and Dehaene's (2012) suggestion about the possible nature of the BP arising as a result of the measurement procedure on an accumulator model impact the timing question here? Should we be reifying the BP, or instead discuss the timing of threshold events? Or both, since the jury is out?

Amber Hopkins

It appears that, at least at present, popular, more global brain-imaging techniques come with a trade-off between higher spatial resolution (fMRI) and higher temporal resolution (EEG, MEG). Would you say this impedes our ability to make progress on pinpointing the neural bases of many phenomena whose generating components fall under the radar with current methods? Are we simply unable to access much of what is going on in the brain? Or does the combination of these methods give us an adequate sense of the dynamics of the brain?

Replies to Follow-Up Questions

Mark Hallett and Aaron Schurger

It is well known that causality is difficult to establish in science, partly because it is difficult to define operationally. In spite of that, neuroscientists do care about causation and are interested in studying causal interactions in neural systems. One important question that often comes up in neuroscience at the systems level is which brain region is driving which other brain region(s). Trying to answer this sort of question necessarily involves making inferences about causality. Neuroscience has developed techniques for inferring causality, like Granger causality and dynamic causal modeling, that are very popular and have been widely used. In addition, one of the main reasons for the popularity of TMS is that it can be used experimentally to infer causality: if interfering with a given brain region has an impact on task performance, then we infer that the brain region in question plays a causal role in performing that task.

One thing most people agree on is that causes must precede their effects in time, as pointed out by Sinnott-Armstrong in his question. This simple constraint allows us to rule out a causal interaction between A and B when A follows B in time. So, for the purposes of ruling out a putative causal relationship, being able to precisely time brain events can definitely be useful. But even then, we still need a precise and well-crafted operational definition of the phenomenon whose causal role we are investigating. To see how challenging this can be, consider the following conception of the will from Hieronymi's chapter: "that collection of more or less ordinary, interacting aspects of the person's psychology (their cares, concerns, beliefs, desires, commitments, fears, etc.) that generates intentional or voluntary or responsible activity."[7] Brain activity that meets these criteria is going to be very difficult to delimit, which is a necessary first step if we are going to time it. Of course, philosophers never promised to generate easy targets for neuroscientists to hit, but this brings to light one potential truth about neuroscience: that the way the brain operates need not map onto the way human observers of human behavior model that behavior. We speak of things like "personality"—a useful concept for describing human behavior— even though such a thing is unlikely to have any clearly delimitable neural correlate. But from a neuroscientific standpoint we will need to identify that correlate if we are to apply the methods that we have at our disposal for timing it.

One important assumption in timing the onset of brain events underlying cognitive phenomena is that there is a precise onset in the first place. This consideration is important in light of Roskies's question about Schurger's stochastic decision model of the BP.[8] According to the model, the BP has no "onset" per se, but rather is analogous to the gradual decrease in atmospheric pressure preceding rainfall; the rainfall itself may have a reasonably precise onset time, but the downward trend in atmospheric pressure preceding it may not. Stated in another way, you can think of the BP as exponentially decaying into the past; it

---

[7] See Chapter 2 by Hieronymi on the will.
[8] See Schurger et al. (2012).

has a characteristic time constant, but there is no moment at which it starts. The only moment that is meaningful in the context of this model is the moment at which the decision threshold is crossed, leading to movement initiation.[9] This is a real event that we could in theory try to pinpoint in time. The jury is still out on whether the stochastic decision model or some other model is the better interpretation of the BP, so at this stage we should be testing different hypotheses and trying to experimentally rule out one model or another.

Additionally, as in the answer to the original question, some events are truly extended over time and not a single point in time. These can still be measured in time with, theoretically, a beginning, a duration, and an end. The required temporal resolution to identify a process extended in time is lower than if it is a moment in time.

Finally, in answer to Hopkins's question, the trade-off between spatial and temporal resolution of different neuroimaging modalities is definitely an impediment to our ability to address questions about the locus of neurocognitive phenomena like "intentions." The greatest impediment is that such phenomena might be widely and sparsely distributed and thus not associated with any single locus in the brain.[10] Some neurocognitive phenomena, like face perception, seem to be associated with specific cortical loci (and even that is debatable), but others, like consciousness, probably cannot be localized to a small region in the brain. The degree to which intentions are localizable in the brain is still a matter of debate. Even if they are localized, it might be difficult to delimit their location non-invasively with techniques like EEG, MEG, or fMRI if high temporal and spatial resolution are needed simultaneously. If intentions are not localized but rather are diffuse and widely distributed, then localizing them would, de facto, not be possible. Rather, we would have to speak of intentions in terms of brain states rather than single brain loci or networks of loci. If, on the other hand, intentions are extended in time, high temporal resolution might not be needed and techniques like fMRI might well be able to be employed. In fact, there is evidence that it is possible to identify brain

---

[9] For more on this, see Chapter 30 by Kreiman et al. on computational models and free will.

[10] See Chapter 22 by Haynes on evidence that intentions are represented in the brain.

states with fMRI and with EEG, in the latter case called microstates. The correlation of such states with conscious phenomena is ongoing work (Bréchet et al., 2019).

However, yes, certain brain phenomena may indeed fall under the radar of current neuroimaging techniques due to the trade-off between spatial and temporal resolution. This might be further complicated by the fact that some phenomena do not localize to a specific brain region or network of regions,[11] but rather might be diffuse and widely distributed, requiring technology that can record from the entire brain simultaneously with very high spatial and temporal resolution. A new technique called functional ultrasound may hold some promise in that regard.

---

[11] For examples, see Hopkins & Maoz's replies in Chapter 23 on the neural correlates of beliefs and desires.

28

How can we determine the precise timing of *mental* events related to action?

Sae Jin Lee, Sook Mun (Alice) Wong, Uri Maoz, and Mark Hallett

Action, to the extent that it involves external, bodily movement, can typically be easily, accurately, and objectively timed. In contrast, mental events are internal and subjective and hence much more difficult to time precisely. Nevertheless, various such mental events are related to action, including beliefs, desires, wishes, wants, urges, and intentions.[1] And yet perception of such internal, mental events is sensitive to errors stemming from inaccuracies, distortions, biases, and illusions, to name a few causes. Here we will focus on what appears to be a critical mental event for action: the sense of willing or intending to move. In typical accounts, an intention must precede movement for it to be experienced as causing the movement. Moreover, one might have a *proximal intention* to act, i.e., an intention to act now (e.g., an intention now to reach for a cup of water), or a *distal intention* to act, i.e., an intention now to act later (e.g., an intention now to go hiking in a few minutes, this weekend, or next year). In the following, we will deal only with proximal intention.

Libet, Gleason, Wright, and Pearl (1983) made the first attempt to time mental events related to action using a clock composed of a rotating dot that completed a rotation on the screen of an oscilloscope every 2.56 s. Participants were asked to note and report the position of the dot when they first became conscious of the urge or the decision

---

[1] For more on beliefs and desires, see Chapter 23 by Hopkins & Maoz. For more on intentions, see Chapter 1 by Yaffe and Chapter 21 by Parés-Pujolràs & Haggard.

Sae Jin Lee, Sook Mun (Alice) Wong, Uri Maoz, and Mark Hallett, *How can we determine the precise timing of mental events related to action?* In: *Free Will*. Edited by: Uri Maoz and Walter Sinnott-Armstrong, Oxford University Press. © Oxford University Press 2022. DOI: 10.1093/oso/9780197572153.003.0028

to move. This reported time—known as W-time—was then compared to brain activity preceding movement, as measured with electroencephalography (EEG), while the decision to move was being made. Similarly to Kornhuber and Deecke (1965), Libet et al. (1983) found an electrical potential could be seen in the brain prior to the actual movement, termed the Bereitschaftspotential, or readiness potential (RP) in English. The RP began about 550 ms before movement onset. Libet and colleagues clocked W-time to about 200 ms prior to movement, suggesting that the RP begins before subjects report being aware of their urge or intention to move. Many investigators have criticized the use of the Libet clock to measure W-time. For example, W-time is retrospectively reported (although it is supposedly contemporaneously encoded into memory). The clock also moves regularly (i.e., the dot rotates at a constant speed), enabling subjects to potentially plan ahead—consciously or unconsciously—and decide to move when it reaches a certain point. Other researchers have since developed various methods to time mental events related to action.

Matsuhashi and Hallett (2008) introduced a novel probe method to measure the onset of the intention to move that does not rely on subjective or retrospective reporting or on clock monitoring. Participants were instructed to carry out brisk, self-paced index finger extensions as soon as the thought of movement came to their mind, without counting or trying to determine the time of the movement, 5 to 10 s after trial onset. During the experiment, auditory tones would play randomly. Subjects were further instructed to veto their movement if they heard a tone and already had an intention to move. Thus, comparing the timings of tones and movements, the investigators found a period before the movements when the tones tended to be vetoed. Using this method, they clocked W-time (called T-time in this experiment) to about 1,400 ms before movement onset, or more than 1 s earlier than Libet's W-time. It is noteworthy that, with more advanced technology, the RP in this experiment was found to begin considerably earlier than in the original Libet study—more than 2 s before movement onset. So in this experiment too, RP began before W-time. The time between T-time and W-time was considered a period of "probe-awareness" characterized by recognition of an intention only if probed.

Bai et al. (2011) and Schneider, Houdayer, Bai, and Hallett (2013) used real-time prediction of movement onset—before it occurred—to measure the relative timing of brain events and intentions. Here subjects were again instructed to carry out self-paced voluntary wrist extensions. When they moved, a light went on. An algorithm was trained on EEG activity during calibration blocks and then tested in real time during prediction blocks. Voluntary movement could often be predicted (albeit with some false positives) around 1.5 to 0.5 s before movement onset, and when predicted, the computer turned the light on. In many of the prediction trials—not surprisingly— the participants moved anyway, but in others, the participants did not. In all trials where the computer predicted movement, subjects were asked what they were thinking when the light went on. In some trials, subjects reported having planned to move; in others they reported having thought of something else. So this study lends evidence that, at least sometimes, movement planning begins unconsciously.

When interpreting the results of all three methods, and especially when comparing intention onset between experimental methods, the literature on time perception raises concerns. For example, clock speed, stimulus intensity, predictability, attention, repetition, and modality all produce systematic distortions in time perception (Allman, Teki, Griffiths, & Meck, 2014; Eagleman, 2008). These factors would have direct implications for interpreting the lengths of intention-onset times across different experiments using different clocks. In particular, we know that our subjective sense of time is highly influenced by how much we conceive the temporal quality of events (i.e., attentional time-sharing). This is key to experimental setups that ask participants to attend to and time internal events like intention onset (Allman et al., 2014).

While perhaps imperfect, these studies, taken together, nevertheless suggest that intention formation might be a process extended over time rather than a momentary event. This process, or a closely related correlate, seems to be identifiable up to 1,400 ms before movement onset, but rises to full consciousness only around 200 ms before movement onset.[2]

---

[2] The authors of this chapter recommend Haggard, Clark, and Kalogeras (2002) for further reading.

Follow-Up Questions

Adina Roskies

When people are asked to report whether or not they had an intention, they are asked to form a mental state whose object is their intention. But this is not a natural state; we do not typically have intentions as our objects of thought, but rather our goals, actions, etc. Do you think this is a plausible interpretation? If so, should we distinguish conscious intention, which we think of as important for voluntary action, from consciousness *of* intention, which is what these studies measure, and which may not be important for voluntary action?

Alfred R. Mele

You say that the studies you discuss "taken together . . . suggest that intention formation might be a process extended over time rather than a momentary event." What's your attitude toward the following idea? The process that causes an intention (in situations of the kind studied) is extended over time, but the intention issued by the process comes into existence at a moment in time. How do the findings you discuss bear on this idea?

Dehua Liang

Self-report of intention onset, as in the Libet clock paradigm,[3] appears problematic. For example, this W-time was shown to move forward in time in line with delayed illusory feedback from movement onset (Banks & Isham, 2009). Are there therefore more objective measures of intention onset—e.g., pupil dilation or eye movement—or perhaps other paradigms that may be less confounded?

---

[3] For more on this paradigm, see Chapter 19 by Bold et al. and Chapter 28 by Lee et al.

Adina Roskies

In the Bai et al. (2011) experiment, in cases in which the light was turned on but the person did not report having had an intention, why is that evidence that intentions begin prior to awareness? Could the algorithm have output a false positive? (How does the false positive rate compare to frequency of light but no report of having intended?)

Replies to Follow-Up Questions

Sae Jin Lee, Sook Mun (Alice) Wong, Uri Maoz, and Mark Hallett

In response to the question posed by Roskies, the word "intention" is used in different ways; here we mean conscious perception of the plan to do something. Ours is the same sense as the word is used in the questions. When people do things, they don't ordinarily think about their intentions; they just move. For example, people generally don't have an intention to talk; they just do. So, we agree, it is plausible that intending and reporting an intention are different, and we should distinguish between the process of intending (whether conscious or unconscious) and consciousness of that process just as we distinguish between cognition and metacognition. The setting of Libet's (1983) experiment is indeed artificial because participants are forced to not only think about the intention but also to time the intention. Theoretically, we should distinguish between (a) the time of the brain event underlying a plan to move, (b) the time of awareness of that plan, and (c) the time that is reported to be the time of awareness (i.e., Libet's W-time). Libet's W-time is a metacognitive report of intention akin to asking someone to recall a list of words and report when they have finished remembering all the words on the list. Matsuhashi and Hallett (2008), however, provide one method to directly access the intention process (point b) without going through the additional layer of metacognitive report, akin to asking a person to write down the words as they remember them, and then measuring the time taken for the person to write the entire list without asking them to report the time themselves.

All three points are interesting, and they may or may not have fixed relationships. The most important for movement would be the brain event, as that drives the motor cortex.

In response to Mele's question, the evidence that brain events are extended over time seems good. For example, the RP is extended over 1 to 2 s. In contrast, qualia seem to just pop into consciousness. There are situations in which the conscious intention to move gets stronger over time. Consider an itch that does not go away. The urge to scratch gets gradually stronger over time. However, the onset of the itch does seem to start at a specific moment. Perhaps the process that causes an intention and the intention itself can be likened to memory retrieval. The process of retrieving a memory is extended in time but leads to a retrieval of memory that can be pinpointed in time; so too can the process of intending lead to an intention that can be pinpointed in time. Here, the term "intending" is potentially confusing, as it usually means explicit intention.[4] We need to be careful with our terms, as Adina Roskies pointed out. But just as the memory retrieval process can have explicit and implicit portions, the intending process may too. The Matsuhashi and Hallett (2008) experiment comes closest to identifying the gradual development of intentions. Tones probed the intention to move and found a much earlier time (called T) than Libet's W for its onset. However, during the time between the Matsuhashi's T and Libet's W, there is no quale of intention, so the probe is of the unconscious brain events that relate to planning for the upcoming movement.

In response to the question from Dehua Liang, the time report of awareness (Libet's W) has been shown to be affected by various perturbations and does not seem to be fully reliable. The methodological advantages of having measures of intending and intention that are not self-report or metacognitive are clear. It would indeed be helpful if we had measures such as the ones mentioned in Liang's question (pupil dilation, eye movement), but we have only recently started to investigate whether those measures are a good index of intending. Physiological measures such as EEG and pupil dilation are

---

[4] See Yaffe's replies in Chapter 1 on intention.

good measures of brain events in general, but their exact relationship to awareness has not been demonstrated. We are also only starting to investigate paradigms that could be used to measure intending and intention. The Matsuhashi and Hallett (2008) experiment was one attempt at a more objective approach, but the authors finally concluded, as noted already, that the first part of the interval between T and W was prior to consciousness, and the exact moment of transition from "probe-awareness" to "meta-awareness" was not defined.

Finally, responding to Roskies's second question, even if the Schneider et al. (2013) experiment was perfect, it would not be a good objective measure of the exact time of awareness in relation to movement onset since, if the experimenter turned on the light, the subject did not necessarily move. Moreover, as noted in the question, there were false positives and false negatives. Considerable attention was paid to this issue in the method. The EEG model was created with data from 240 trials prior to the experiments and checked again after the experiment with 60 trials. False negatives were not a big concern, but false positives were. The false positive rate for predictions was kept at 18%. In the experiment 32% of predictions were without intention, and thus about 14% were true positives, but it could not be determined which were true and which were false. The experiment could conclude only that some of the time movement could be predicted when the subject had no intention to move. And, therefore, that is the sense in which the experiment provides evidence that intentions sometimes began prior to awareness.

29

Are any neural processes truly random (or stochastic)?

Hans Liljenström

Living systems are normally associated with a high degree of order and organization. However, disorder of various kinds is also abundant in both structure and behavior. In particular, neural systems and their processes are characterized by a large degree of random "noise," or fluctuations at different levels (see, e.g., Århem, Blomberg, & Liljenström, 2000; Destexhe & Rudolph-Lilith, 2012). Whether these fluctuations are of stochastic (i.e., indeterministic) or deterministic origin is debatable.[1] In fact, a deterministic, high-dimensional, chaotic process (time series) and a purely stochastic process would be so much alike that it is practically impossible to distinguish between them. Still, electrophysiologically observed fluctuations at microscopic levels are generally considered indeterministic, primarily resulting from the random motion of molecules. Fluctuations at macroscopic levels are instead mostly regarded as deterministic chaotic, due to the complex interaction between various excitatory and inhibitory neurons in intricate neuronal networks. (The stochastic events at microscopic levels are supposed to be averaged out at macroscopic levels.)

A time course of a strictly deterministic process is entirely determined by its initial conditions, and its future development can occur only in a unique way, according to some pre-set rules, as given, for example, by natural laws. Hence, for such processes, knowing the state of the system at some point in time and those rules of development, one can—in theory, though not always in practice—compute all its future and past states. A deterministic chaotic process (*chaos*) is typically

---

[1] For more on determinism, see Chapter 6 by O'Connor.

Hans Liljenström, *Are any neural processes truly random (or stochastic)?* In: *Free Will*. Edited by: Uri Maoz and Walter Sinnott-Armstrong, Oxford University Press. © Oxford University Press 2022.
DOI: 10.1093/oso/9780197572153.003.0029

extremely sensitive to initial conditions, implying that the time course is different every time the process starts. Hence, this process is not predictable at longer time scales, while it could be predictable at shorter time scales. In contrast, a stochastic process is a collection of random variables, or a system which evolves in time while undergoing chance fluctuations, such as in radioactive decay or the motion of gas particles. The key feature that distinguishes such systems from deterministic ones is that even if its current state and the rules governing its development are completely known, the next state is still not predictable. Such indeterministic processes are typically termed *noise*. A distinction between noise and chaos could be described in terms of mechanisms and simplicity: chaos originates from a simple, controllable mechanism (which could be described by a mathematical equation), while noise originates from a large number of uncontrollable mechanisms. Both types are examples of unpredictable fluctuations in neural systems.

At all neural levels, irregular fluctuations are at work. At a *microscopic* level, the motion and conformation changes of atoms and molecules result in stochastic processes as thermal noise, or so-called Brownian motion. Some argue that *true* randomness is found only in the quantum world of atoms and possibly of smaller molecules, and indeed, quantum effects may be involved in the stochastic flux of ions through the neuronal ion channels (membrane proteins). However, for all practical purposes, the random opening and closing of ion channels is generally considered to be stochastic due to thermal noise as the main contributor to the observed *ion channel noise*.

At a *mesoscopic* (intermediate) level, the typical firing pattern of cortical neurons is highly irregular, whether stimulated by sensory input or appearing as spontaneous firing of action potentials (AP) in the absence of external stimuli. A substantial part of the (apparent) stochastic neuronal firing originates from ion channel noise, but another major source is *synaptic noise*, i.e., the probabilistic nature of synaptic transmission. Individual synapses appear to be highly stochastic and "unreliable," releasing neurotransmitter substance in response to as few as 10% of the APs arriving at the synaptic cleft.[2] There is also a

---

[2] That is, only 1 out of 10 APs arriving at the synapse will lead to a release of transmitter substance, and hence to a signal transfer.

probabilistic uptake of neurotransmitters at the receiving end of the synapse. Since many neurons have only a few synaptic connections with any other neuron, this may result in quite low reliability of signal transmission from one neuron to the next.

While the behavior of single molecules or cells may appear stochastic and noisy at micro- and mesoscopic levels, their collective behavior could generally be regarded as deterministic and ordered at the *macroscopic* level, where the irregularities at lower levels are "averaged out."[3] For example, the stochastic openings and closings of thousands of ion channels may result in the firing of an AP, and the irregular firing of thousands of neurons may be functionally efficient in signal transmission across brain regions (Freeman, 1996). (Even the daily loss of thousands of neurons does not normally lead to significant impairment of our cognitive or conscious processes).

Despite the general order at the macroscopic level, fluctuations may also exist in cortical networks, as a result of either the spontaneous firing of neurons or the interaction of a large set of interconnected excitatory and inhibitory neurons. The resulting complex neurodynamics observed with, e.g., electroencephalography or other brain-imaging techniques can be characterized as both noisy and chaotic, in addition to oscillations at various frequencies.[4]

As demonstrated, a certain degree of disorder seems inevitable at every level of neural organization, and organisms have presumably evolved to cope with it. The question is whether it is also advantageous to the system, as evidence suggests (Århem, Blomberg, & Liljenström, 2000). Could the irregular neural activity, described as noise or chaos, play any functional role in the nervous system, with significance for any sensory, cognitive, or conscious processes?

Ordered signals can be used for computations and prediction, but all of the described neuronal disorder appears to make exact predictability impossible and any computational process difficult. However, the spontaneous irregular firing of neurons could be important for maintaining baseline activity, necessary for neural survival and/or for

---

[3] The reason for this is somewhat similar to the fact that a rock appears solid and stationary, while its constituent atoms are mostly empty space and in constant motion.

[4] For more on neuroimaging, see Chapter 27 by Hallett & Schurger.

readiness to respond to input. Internal, system-generated fluctuations can also create state transitions, breaking down one kind of order (e.g., oscillations at some frequency) and replacing it with a new kind of order. Externally generated fluctuations can cause increased sensitivity in receptor cells through the phenomenon of *stochastic resonance*, in which weak signals can be amplified by noise. More specifically, the signal-to-noise ratio can increase for increasing noise levels and reach a maximum for an optimal value of the noise. Experimental evidence for this phenomenon in biological systems has been found in, e.g., sensory neurons of crayfish and various kinds of fish, but also in mammals (see, e.g., Wiesenfeld & Moss, 1995).

In addition, computer simulations demonstrate that system stability and flexibility depend on the level of intrinsic noise, as well as on (synaptic) connection strength and neuronal excitability. With proper regulation, the complex neurodynamics can keep the system at an appropriate balance between flexibility and stability. Insignificant fluctuations can be ignored, while small signals can be amplified and result in an adaptive system response. Microscopic fluctuations can result in global oscillations and state transitions at a network level. In particular, the rate of information processing in associative-memory tasks can be maximized for optimal noise levels. For example, the system may recognize a learned pattern faster with some degree of noise in the input signal than without any noise at all. Too much noise in the input signal may, however, make it harder or even impossible to associate that signal with any learned input patterns. Noise can also induce transitions between different dynamical states that could be related to different functional states of perception, learning, and memory (Århem, Blomberg & Liljenström, 2000; Århem, Braun, Huber, & Liljenström, 2005).

To conclude, neural systems and processes are characterized by a high degree of disorder at different levels, from stochastic processes of molecules and neurons to the complex brain dynamics observed with, e.g., electroencephalography. The bottom line is that both indeterministic and deterministic processes can account for the observed (unpredictable) fluctuations in neural processes. Whether this neural unpredictability is sufficient to account for free will is not (yet) possible to determine experimentally or otherwise.

Follow-Up Questions

Timothy O'Connor

You note the familiar point that stochastic events at micro and meso scales might be averaged out to virtual determinism at the macro scale. Might you comment on the extent to which it remains an open question whether this holds true for every significant neural function?

Amber Hopkins

In Timothy O'Connor's answer to the question "Can there be free will in a determined universe?,"[5] he states, "[F]ree will is a paradigmatic system-level or 'top down' form of control. If we are free, it must be that we ourselves determine which choices we make." You state that "microscopic fluctuations can result in global oscillations and state transitions at a network level." Is there any evidence of these macroscopic or global oscillations impacting microscopic fluctuations in a top-down manner?

Adina Roskies

Could you say more here about the relevance of unpredictability to the question of free will: indeterminism is necessary for incompatibilist free will, but unpredictability does not equal (metaphysical) indeterminism. However, there may be notions of free will for which unpredictability itself matters, even if they are deterministic (Roskies, 2006).

Replies to Follow-Up Questions

Hans Liljenström

My original response primarily concerned the generation and origin of neural fluctuations at different scales but didn't say much about the

---

[5] See Chapter 6 by O'Connor on determinism.

relevance to free will, as the many follow-up questions indicated. For example, Timothy O'Connor wants to know if stochastic events at micro and meso scales might be averaged out to virtual determinism at the macro scale for every significant neural function.

Presumably, most, if not all, neural functions, including our cognitive and conscious processes, require a large number of neurons firing, and the fluctuations at molecular and neuronal levels are typically smeared out at the level of cortical networks.[6] In some special cases, events at microscopic levels, such as the retinal absorption of single photons under extremely dark conditions, or single APs traveling from a stimulated finger to the brain, can be amplified by various neural networks and give rise to perceptions. There is even evidence that the opening of single ion channels may elicit APs in small hippocampal interneurons, resulting in spontaneous firing of those neurons. With a computational model of hippocampus, we have demonstrated how such microscopic fluctuations can result in phase shifts in the complex neurodynamics at the network level, with accompanying shifts in system functions (Århem & Liljenström, 2001). So even if the micro- and mesoscopic fluctuations, in general, are averaged out, there may be cases where they are not. The sensitivity and capacity to amplify microscopic events appears to be under neuromodulatory control, an effect we have also simulated with our computational models.

Neuromodulators, such as acetylcholine and serotonin, can change the excitability of a large number of neurons simultaneously, as well as the synaptic transmission between them. The concentration of these neuromodulators is related to the arousal/attention/motivation of the individual and can have profound effects on cortical neurodynamics and on cognitive functions, such as associative memory (Liljenström & Hasselmo, 1995). This very likely also effects decision-making, where the integration of various sensory and intrinsic signals would require delicate regulation (Nazir & Liljenström, 2015). In addition, computer simulations of cortical networks have demonstrated how the complex neurodynamics (oscillations and chaos) itself can entrain and synchronize the activity of large populations of neurons, which initially were silent or firing randomly (Liljenström, 2016). I believe these

---

[6] See, e.g., *Mass Action in the Nervous System* by Freeman (1975).

examples could be a response to Amber Hopkins's question regarding macroscopic neurodynamics impacting microscopic fluctuations in a top-down manner.

In general, it is quite clear that higher levels in biological systems exert their influence over lower levels.[7] Each level provides the boundary conditions under which the processes at lower levels operate. Without boundary conditions, biological functions would not exist.[8] In particular, the intricate web of inter-relationships between different organizational levels of neural systems may provide both upward and downward causation. A theoretical basis for this is given by *synergetics* (Haken, 1991); Freeman also relates to this when describing the *circular causality* in the action-perception cycle of an individual.[9] Hence, mental processes may arise from neural activity, but they may also affect that activity, in a kind of downward (or circular) causation that seems necessary for free will.

But could the neural unpredictability that we have discussed say anything about free will, as Adina Roskies wonders? An assumed spontaneous volitional movement of a limb might be unpredictable, but that unpredictable movement could also have resulted from some stochastic or chaotic neural process. Our current scientific methods may not (yet) be able to determine the cause of the movement (except when artificially manipulating the nervous system). Also, theoretically, neither the indeterminism of stochastic processes nor the (alleged) determinism of natural laws seems to allow for free will as an explanation.

Perhaps the main reason why science has problems encompassing free will is that it seems to entail causal agency, requiring the action of a conscious agent, and (natural) science has so far dealt only with objects, not agents/subjects. However, the theories and laws of physics, including that of determinism, were developed for comparatively simple physical objects and their interactions. They have very little

---

[7] See Chapter 20 by Hallett on levels and control in the brain.

[8] This has been recognized by many theoretical biologists, such as Robert Rosen and Stuart Kaufmann, but perhaps best expressed by Denis Noble (2008) in his book *The Music of Life*.

[9] This is described in several publications by W. J. Freeman, perhaps most extensively in Freeman (2000), but also in Liljenström (2018).

to say about the behavior of complex biological systems,[10] in particular brain-mind systems and their (inter-)actions. In order to allow for consciousness and free will, science probably needs to be extended beyond chance and necessity, which currently are its only models of explanation.[11]

---

[10] For example, Erwin Schrödinger (1944, p. 76) states in his groundbreaking book, *What Is Life?*, "From all we have learnt about the structure of living matter, we must be prepared to find it working in a manner that cannot be reduced to the ordinary laws of physics."

[11] See, e.g., *Chance and Necessity: An Essay on the Natural Philosophy of Modern Biology* by Nobel laureate Jacques Monod (1971). See also Chapter 26 by Hopkins et al. on consensus on consciousness and volition.

30

How can computational models help us understand free will?

Gabriel Kreiman, Hans Liljenström, Aaron Schurger,
and Uri Maoz

Richard Feynman famously stated, "What I cannot create, I do not understand." And in that spirit, understanding the relations between volition and action (or any other aspect of brain function) would likely benefit from building *quantitative, falsifiable, predictive, computational models* of those relations. Verbal models that are not falsifiable and do not make quantitative predictions can be interesting and inspiring but do not satisfy the basic requirements of scientific models.

The desiderata for a computational model with relevance to free will include being able to account for the critical features of volition—leading to movement, being endogenously triggered, responsive to reasons, goal directed, spontaneous or innovative, and involving consciousness—as articulated by Patrick Haggard (2019). Importantly, the model should quantitatively predict neural activity in relevant brain circuitry during free-will or volitional tasks, and also predict the behavioral responses of the agent during volition tasks. Prediction should be understood within the context of noise in the system given available measurement techniques. As an analogy, we have a reasonable understanding of fluid dynamics and statistical mechanics, yet it remains daunting to predict the trajectory of every single molecule in the air. Of course, the deterministic chaotic aspects of such a system (and in particular its extreme sensitivity to initial conditions)[1] and the noise in our measurements should not be interpreted to imply that

---

[1] See Chapter 29 by Liljenström on stochastic neural processes and Chapter 6 by O'Connor on determinism.

Gabriel Kreiman, Hans Liljenström, Aaron Schurger, and Uri Maoz, *How can computational models help us understand free will?* In: *Free Will.* Edited by: Uri Maoz and Walter Sinnott-Armstrong, Oxford University Press. © Oxford University Press 2022. DOI: 10.1093/oso/9780197572153.003.0030

gas molecules have any form of will that is unaccounted for by current models. It is perhaps also worth emphasizing that a computational model of free will would not have free will any more than a computational model of black holes would have gravity.

Advances in computer capacity and algorithms, together with the tremendous growth of data in neuroscience, have dramatically improved the possibilities of more realistically modeling and simulating certain brain structures and activities. Hence, there is growing confidence in our ability to use models and simulations to gain deeper insights into the functioning of the human brain, including cognitive processes like associative memory, pattern recognition, and motor control. Computer simulations should be viewed as essential to guiding and interpreting experiments in our pursuit to unravel the mysteries of volitional decisions.

There have not been many computational models of free will. Existing efforts may build on common models of decision-making and often rely on variations of integration to a threshold, also referred to as drift-diffusion models (described later). In a typical decision-making context, the model would incorporate sensory evidence that figures prominently in the decision process. However, in a "free will" task, the decision is not determined solely on the basis of sensory evidence. Hence, threshold crossing must be driven by other factors—such as internal variables. Consider a binary decision task, e.g., "Decide whether to raise your right hand or your left hand." The final effectors involve motor cortex mechanisms that culminate in motor-neuron activity. Such high-level models posit that the input to those motor effectors involves a competition between populations of neurons that push for "left" or "right." This activity might be sub-threshold from the point of view of the conscious sensation of will. According to this family of models, this sub-threshold activity accumulates over time until it reaches one of the two thresholds, reflecting the selection of one of the two decision alternatives—in this case moving the left or right hand (see Figure 30.1; extensions of this model beyond binary choices exist).

Mathematically, the drift-diffusion model can be described by:

$$V_t - V_{t-\Delta t} = D\,\Delta t + c\;sqrt(\Delta t)\;\varepsilon_t, \qquad (1)$$

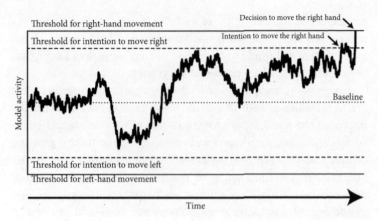

Figure 30.1 Illustration of the drift-diffusion described by Equation 1. According to this model, the intention to move right takes place when activity reaches a threshold (top/bottom dashed lines). A second threshold, potentially in a different brain area, needs to be crossed for the movement to be executed (top/bottom solid lines).

where V_t can be thought of as a decision-voltage variable that is being accumulated; D is a drift rate or urgency to move that may include biases toward one or the other decision alternatives, and which may also include a leak term, instead being $(D-kV)$, whereby the decision decays with time, if it does not reach a threshold; c is a scaling constant; and ε_t is the "noise" term governing the stochastic diffusion process (see Figure 30.1). The noise term shows relatively long autocorrelations that can be described by a power law (Schurger, 2018).

In the model, the *intention*[2] to move might emerge once this activity reaches a threshold before the decision takes place (dashed lines in Figure 30.1). The model can be readily implemented at the neuronal spike level and captures essential properties of the neurophysiological correlates of volitional decisions (Fried, Mukamel, & Kreiman, 2011). One phenomenon that might factor into the neural antecedents of spontaneous voluntary actions, and that could be incorporated

[2] For more on intention, see Chapter 1 by Yaffe and Chapter 21 by Parés-Pujolràs & Haggard.

into drift-diffusion models, is motor rehearsal. Regularly recurring rehearsal of the to-be-performed movement (or one among them) could interact with sub-threshold fluctuations to bring the system over threshold at a given moment that is determined by neither rehearsal nor sub-threshold fluctuations alone. Further work in this area could be very fruitful.

In addition, neural network models of specific cortical areas thought to be involved in decision-making could help us better understand the mechanisms of volitional decisions. A better-defined commitment to specific brain areas can bring these models closer to the neural substrates. The simulated neural activity could be compared to experimental data from invasive neurophysiological recordings or from noninvasive measurements. The neural information processing and the neurodynamics of such areas could be modeled with, e.g., attractor dynamics, using continuous input-output relations. A classic example of attractor dynamics is the Hopfield network, where units are connected in an all-to-all fashion via symmetric weights. In a more biologically realistic model, the connections are aimed at mimicking the structure of any relevant brain area, resulting in more realistic neurodynamics. The state of the network is characterized by multiple attractors, to which the state converges from any arbitrary initial condition, depending on extrinsic or intrinsic input signals. Different options in a decision-making task can then be represented by the oscillatory activity of Hebbian cell assemblies (see, e.g., Nazir & Liljenström, 2015). This kind of neurocomputational models can, for example, be used to simulate and investigate how information transfer occurs in various interacting brain areas during tasks involving volitional decisions.

Both integration-to-threshold models and attractor dynamic models, once activity is settled, convey their signals to motor effectors for execution. But this need not be the case. Veto signals could interfere with execution or might result from a distal decision, not directly involving any motor output. How such veto or modulatory mechanisms would interact with integration-to-threshold models remains unclear.

The integration-to-threshold family of models and dynamic attractor–based models may encompass volitional decisions as well as sensory-based decisions. They usually make no fundamental distinction between decisions where the alternatives are equally valued or not, or situations where those decisions can be influenced by sensory inputs, task goals, or expected rewards. These important variables can influence volitional decisions as well as modulate and provide inputs to the integration processes.

Opinions diverge regarding the extent to which computational models will shed light on free will. Some of the authors argue that computational models can help elucidate the critical mechanisms underlying free will as long as they can quantitatively explain experimental observations at the behavioral[3] and neurophysiological[4] levels. Other authors argue that, if free will requires consciousness, we need to know how to represent consciousness in our computational models before we can use them to explain free will. The computational models might at least account for the emergence of the conscious experience of the intention to move and its relation to the onset of movement, even if the true nature of consciousness remains to be elucidated. Time will tell.

Follow-Up Questions

Walter Sinnott-Armstrong

When a computational model predicts human decisions, how can we know whether or not the computations in that model correspond to parts of a mechanism or process in the human brain? Isn't it possible that the brain operates in a very different way from a model that predicts its output? If so, does the computational model really explain anything about real flesh-and-blood human brains?

Jake Gavenas

Your point that a computational model of free will would itself have free will no more than a model of a black hole would itself have gravity makes sense. However, with a computational model of free will we

---

[3] See Chapter 10 by Bayne on behavioral experiments.
[4] See Chapter 27 by Hallett & Schurger on neuroimaging.

could potentially create an algorithm or robot to act in accordance with said model. To what extent would this hypothetical "creature" have free will?

Amber Hopkins

Models are an abstraction of neurophysiological activity but essential to learning more about the brain and may successfully allow us to make impressive predictions about actions and behavior. Do you find that models match well to what we know about how the brain works through different imaging or measurement techniques?[5] Or do you find that the support for models comes from our ability to derive successful predictions from them rather than how well they match information flow or activity in the brain? It could be that with current techniques we simply do not have access to the processes in the brain that we would need to identify to say a model accurately matches brain function. For example, regarding attractor models, you explain that a system converges on a state from an initial condition, depending on extrinsic or intrinsic input signals. Are intrinsic inputs here volitional percepts, goals, beliefs, and so on? The neural bases of these states or processes are elusive. And the output are motor outputs? This seems clearer/agreed upon. Are input-output relations clear when we measure and record brain activity? Is what the inputs go into and what outputs come out from in the model clear?

Replies to Follow-Up Questions

Gabriel Kreiman, Hans Liljenström, Aaron Schurger, and Uri Maoz

In response to Sinnott-Armstrong's follow-up question, we point out that when constructing a computational model of any neural

---

[5] See Chapter 27 by Hallett & Schurger for more on neuroimaging.

system, we have to make tremendous simplifications. We often apply Occam's razor: a model should be as simple as possible, but not simpler. The problem is to find an adequate simplification, where the essential details are included, while the less relevant ones are neglected. Naturally, important features may be missed in that process. Due to this fact, a model is never correct; it is only more or less useful depending on the purpose. A model of any brain structure/process can be useful only to the extent it can describe plausible links between structure, dynamics, and function or predict the outcome of an experiment. Just simulating a neural process or a cognitive function can never give us any conclusive understanding of the system or process being modeled; it can just demonstrate possible solutions based on known facts and assumptions made.

Depending on its purpose, the model may include more details at a particular organizational level, e.g., cortical networks involved in decision-making, while excluding details at other, e.g., molecular levels. In the same way, a modeled structure may only crudely correspond to the biological origin, while the dynamics and functions could be similar. The famous Hodgkin-Huxley model of action potential generation is an example of a success story (i.e., extensively used), where all the intricate molecular and ionic details are lumped into mathematical equations with a few parameters that can be experimentally determined. Another example is the Hopfield network of associative memory, where populations of neurons are modeled with simple input-output relations and all-to-all connections, which is far from known facts about real neural networks but nevertheless gives a (partial) understanding of how such networks may store memories. A good model can be helpful in understanding a system or process, predict the outcome of an experiment, or guide new experimental studies.

Responding to Gavenas, we note again that a model is necessarily a simplification of the real phenomenon that it is trying to describe and is as valuable as it is useful, rather than expecting it to be a faithful representation of *all* aspects of that phenomenon. But if a system could be constructed, based on a certain model, that possesses all of the necessary and sufficient conditions for free will, according to some theory, then we would have to concede that it does have free will, assuming that the theory is correct. This is what it means to commit to a theory.

In your question you asserted that "with a computational model of free will we could potentially create an algorithm or robot to act in accordance with said model," as if to suggest that the same would not be logically possible with the example of a black hole. But that is not necessarily the case. Supposing that you took a mathematical model of a black hole and tried to construct something that had all of the properties that were dictated by those equations (imagining, for the sake of argument, that you could actually do that). Depending on how faithful your model was, you might end up creating an actual black hole (and, should you do so from within the event horizon, no longer be around to tell anyone about it)! A black hole is a well-defined concept, and the existence of black holes was predicted by general relativity well before signs of their actual existence were ever observed. By contrast, there is no model, or at least no consensual model, from which free will emerges as a predicted property. Imagine, e.g., a hypothetical neural model of the brain that, when you work out all of the math, predicts the existence of free will. It is unclear whether such a model could even exist, but it would depend critically on how we choose to define free will. So maybe we neuroscientists are going about it all wrong. Instead of starting with the concept of free will and trying to construct a model that accounts for it, maybe we should be trying to construct a faithful neural model of human behavior and asking whether or not the existence of free will (however we may choose to define it) is predicted by that model. If we could account for all of the measurable physical facts without any such property emerging as a necessary consequence, then this would cast doubt on the scientific utility of the notion of free will. However, if such a property reliably emerges from models of a certain kind of agent, then that would count as evidence in favor of the existence of free will (again depending on how we choose to define it).[6]

As a response to Hopkins, let us begin with an aspiring example of how models can help us quantitatively understand and predict neurophysiological responses. In the visual system, deep convolutional neural networks provide an initial approximation to the cascade of processing steps along ventral visual cortex (Kreiman, 2021). These

---

[6] See Chapter 9 by Roskies on neuroscientific evidence about free will.

are image-computable models (meaning that the input to the model is an image, as opposed to verbal models that cannot be computed on) that can perform the same experiment as the monkey or human and produce activation patterns that explain about half of the variance in neuronal firing rates. Recently, an attractor network has been incorporated into these architectures to model the process of pattern completion (Tang et al., 2018). But we are interested in modeling volitional decisions! In the most pragmatic version, the input to the model should be the same as the input to the subjects (instructions, a rotating clock, two images to choose from, etc.). Often, such inputs could be replaced by abstractions (such as a scalar variable that reflects the passage of time or a variable that dictates whether a task is cued or not). The output should be measurable variables (such as firing rates or the right index finger pressing a button or moving a clock's hand). Goals, beliefs, and percepts could be indirectly assessed through adequately designed experiments that lead to behavioral outputs. For example, a model unit could represent movement of the right hand and the input to that unit could be interpreted as a proxy for the percept of intention. In sum, inputs to the model should be the same as the inputs to the subjects (modulo abstraction and recodification), and model outputs should be the same as the subjects' behavioral measurements (modulo abstraction and recodification). To be clear, we do not have such models yet, and so this is an aspirational goal rather than a description of existing theories or models.

Glossary of Crucial Terms

Claire Simmons and Amber Hopkins

abulia – A disorder of diminished motivation manifesting in a reduction in purposeful movement, speech, social interactions, and so on. Compare *akinesia*.

access consciousness – A person has access consciousness of a bit of information if that information is directly accessible and available globally for reasoning and other kinds of cognitive processing, so that person is able to use that information in a wide variety of cognitive activities. Contrast *phenomenal consciousness*. See Chapter 12.

action/act – See Chapters 3, 14, 16, and 21.

action potential – The change in electrical potential resulting from the passage of an electric impulse inside the neuron's axon. As a result of the action potential, the neuron releases neurotransmitters and thus sends information to other neurons to which it is connected via synapses. Action potentials drive the primary form of communication between neurons. See Figure 4 in the Brain Maps section.

activity (brain, neural, or neuronal) – Generally refers to electrical or neurochemical changes within or between neurons in the brain. The term is more typically used to refer to synchronized activity of many neurons that can be measured via neuroimaging techniques. See *action potential, neuroimaging*, and Figure 4 in the Brain Maps section.

afferent (signal) – A signal conducting or conducted inward or toward something (for nerves, to the central nervous system). Contrast *efferent*.

agency – The property of performing or being able to perform actions. See *sense of agency*.

agent – A human or other animal that performs an action or has an active role in producing a specific effect. Agents in this context should not be confused with secret agents or chemical agents.

akinesia – A movement disorder manifesting in inability (or impairment in the ability) to initiate voluntary movement. Compare *abulia*.

akrasia – Acting or being disposed or likely to act against one's better judgment due to lack of willpower or self-control. Sometimes described as weakness of will or incontinence.

algorithm – A sequence of well-defined instructions meant to solve a problem, especially by a computer.

anterior – See Figures 1a and 1b in the Brain Maps section.

apraxia – A neurological disorder characterized by the inability to carry out familiar, skilled movements despite the relevant muscles and peripheral nerves as well as the will to move all remaining intact.

arbitrary decisions – Decisions that cannot be made on the basis of reasons for or against the decision alternatives. Making arbitrary decisions is sometimes described as "picking," as opposed to making deliberate decisions, which is termed "choosing." See Chapter 19.

artificial intelligence – See Chapter 11.

ascending (pathway) – A neural pathway that goes up the spinal cord toward the brain and typically transmits sensory information from the body. Contrast *descending*.

attend/attention – A term denoting a variety of functions, all related to focusing on some phenomena and filtering out others. Attention enables the flexible control of limited computational resources.

aware/awareness – The words "awareness" and "consciousness" are often used almost interchangeably. See Chapters 12, 13, 24, 25, and 26.

axon – See Figure 4 in the Brain Maps section.

behavioral – Of or relating to behavior. Behavioral experiments are experiments that measure reaction time, accuracy, and other features of reports and bodily movements outside of the brain. Behavioral data is data provided by measures of such behaviors. See *experiment* and *neuroimaging*.

belief/believe – See Chapters 1, 15, and 23.

Bereitschaftspotential – See *readiness potential*.

bilateral – Relating to both sides. For example, simultaneous activation of the two brain hemispheres would be termed bilateral activation. See Figure 1c in the Brain Maps section. Contrast *contralateral* and *ipsilateral*.

bottom-up – Moving from a lower level to a higher level. In research on decisions and consciousness, the term usually refers to information processing that starts with concrete information about individual events or experiences and results in abstract or conceptual information about general kinds of events or experiences. Contrast *top-down*.

BP (BP1, BP2) – See *readiness potential*.

Brodmann area – A region of cortex categorized by the structure and organization of cells. These areas were originally defined and numbered by the German anatomist Korbinian Brodmann in 1909.

Brownian motion – Erratic, random movement of microscopic particles suspended in a liquid or gas due to collisions with molecules from the surrounding medium.

caudal – See Figures 1a and 1b in the Brain Maps section.

causal role – The function of something as part of a process that causes an effect.

cause/causation – To say that one event causes another is to say that the cause makes a difference to whether the effect occurs and, on some views, does so by transferring energy or some other quantity from the cause to the effect. Contrast mere correlation or co-occurrence of two kinds of events. Causation can be probabilistic or deterministic. See *determinism*.

central nervous system (CNS) – Part of the nervous system consisting of the brain and spinal cord. See Figure 1 in the Brain Maps section.

chaos/chaotic – Chaotic behavior is exhibited by nonlinear dynamical systems that *appear* to transition between states randomly. Chaotic behavior is so sensitive to its initial conditions that predicting the behavior far into the future is practically impossible. See *random, determinism*, and Chapter 29.

choice – The words "choice" and "decision" are often used almost interchangeably. See *decision*.

choosing – The terms "choosing" and "deciding" are often used almost interchangeably, but sometimes choosing is contrasted with picking arbitrarily. See *arbitrary decisions*.

cognitive control – See Chapter 23.

compatibilism – The claim that determinism is consistent or compatible with free will and/or moral responsibility, so it is possible for agents to be free and/or responsible for acts that are determined. See Chapters 4, 5, 6, 7, 9, and 14.

conscious/consciousness – (a) To say that someone is conscious or has consciousness globally is only to say that they are sensing or thinking something as opposed to being asleep, under global anesthesia, or in a coma or some other state in which they do not sense or think anything. Even if someone is conscious in this global sense, they might not be conscious of some bits of information in their environment. (b) To say that someone is conscious of a particular bit of information is to say either that they have a phenomenal experience with that information as its content or that they have access to that information and can use it in cognitive processing. See *phenomenal consciousness* and *access consciousness*. See also Chapters 12, 13, 24, and 25.

contralateral – Relates to the other or opposite side. For example, the motor cortex in each hemisphere mainly controls contralateral muscles, so the motor cortex in the left hemisphere mostly controls muscles on the right side of the body, and vice versa. See Figure 1c in the Brain Maps section. Contrast *ipsilateral* and *bilateral*, and see Chapter 16.

control – The power to determine or at least influence something. A person controls something when it is within their power to bring or influence it toward a state of their choice. See *motor control, self-control*, and Chapters 2, 14, 20, and 25.

cortex – See Figure 1 in the Brain Maps section.

cued-movement task – See *instructed-movement task* and Chapter 22.

data – Information from a measure of some variable. See *experiment* and *behavioral*.

decide/decision – Making a (distal) decision to perform an action in the future is often understood simply as forming an intention to do that act in the future. In contrast, making a (proximal) decision to perform an action at the present moment is often understood simply as exerting control in order to bring about that action. To say that someone decides that a claim is true means that the person decided to believe that claim or to act as if it is true. It is controversial whether all decisions are conscious. The words "decision" and "choice" are often used almost interchangeably. See Chapters 12 and 16–20.

decode – In neuroscience, decoding often refers to the use of machine learning or other techniques to infer some function of an organism of interest (e.g., behavior) based on patterns in brain activity. With respect to memory, decoding typically refers to retrieving something from memory. Contrast *encode*. See *brain activity*.

deep – See Figure 1b in the Brain Maps section.

deliberation – Considering and weighing various reasons that might favor a decision or its alternatives or considering and evaluating the best means toward goals. Deliberate or deliberative decisions are based on deliberation. Decisions that are based on deliberation are still distinct from and come after the deliberation, which results in the decision. It is controversial whether all deliberation must be conscious. See Chapter 19.

descending (pathway) – A neural pathway that transmits motor control signals from the brain to the spinal cord to control bodily movement. Contrast *ascending*. See *motor control* and Figure 3 in the Brain Maps section.

desire – A mental state of wanting something or wanting something to happen. See Chapters 1, 15, and 23.

determinism – The thesis that a complete statement of the laws of nature together with a complete description of the state of the universe at any time logically entails a complete description of the state of the universe at any other time, so any complete state plus laws guarantee that every event, including every action, will occur instead of any incompatible alternative. Determinism denies that the fundamental laws of nature are statistical (also termed "probabilistic" or "stochastic"). See Chapters 6, 9, and 29.

distal – See *decide/decision* and Chapters 21, 22, 25 and 28.

dopamine – A neurotransmitter that has been associated with reward, motivation, and motor control. See Chapters 20 and 23, along with Figure 4 in the Brain Maps section.

dorsal – See Figures 1a, 1b, and 1c in the Brain Maps section.

dualism – The thesis that minds or mental states, events, or properties are not reducible or identical to physical states, events, or properties. Substance dualism claims that minds are non-physical substances, whereas property dualism claims only that mental properties are not reducible to physical properties, even if they are both properties of the same physical thing. Cartesian dualism is the strong version of substance dualism held by René Descartes. See *emergentism.*

ECoG – See *electrocorticography.*

EEG – See *electroencephalography.*

efferent (signal) – A signal conducted or conducting outward or away from something (for nerves, away from the central nervous system). Contrast *afferent.*

electrocorticography (ECoG) – An invasive brain-recording technique that captures aggregate activity by placing electrodes (small conductors that send or receive electricity) directly on the surface of cortex. It has high temporal resolution as well as higher spatial resolution and better signal quality than electroencephalography (EEG). See Chapter 27.

electroencephalography (EEG) – A non-invasive neuroimaging technique that records electrical activity from the brain using electrodes (small conductors that send or receive electricity) placed on the scalp. EEG has high (sub-millisecond) temporal resolution, but it can typically record only cortical activity, tends to be noisy, and suffers from low spatial resolution (compare to other neuroimaging techniques like ECoG, MEG, and fMRI). See Chapter 27.

emergent/emergentism – An emergent property is a property of a complex system that is not a property of its parts but still depends on the existence of those parts. For example, an orchestra has a distinct sound that is not possible without the participation of each instrument in the orchestra, but no single instrument alone sounds like an orchestra. In the case of minds, to say that a mental property emerges from a complex physical system (such as the brain) is to say that the mental property cannot occur without that physical system, but no part of the system (such as a neuron) by itself has the mental property. No single neuron feels fear. Emergentism about the mind is the claim that all mental properties emerge from parts of the physical world, including but perhaps not limited to the brain. Emergentists often add that a mental property must occur whenever certain physical states or events that realize the mental state occur, but the mental property is still not reducible to its physical realizers. See *dualism*.

empirical – Related to or able to be tested by observations or experiments. See *experiment* and *falsifiable*.

encode – In neuroscience, encoding generally relates to representing something in the brain. In the context of memory, encoding typically refers to storing something in memory. Contrast *decode*.

endogenous – See Chapter 3.

epiphenomenal – To say that a mental state or event is epiphenomenal is to say that it does not cause any physical event, including any bodily movement, though the mental state or event still might be caused by physical events.

Brain Maps
Amber Hopkins and Natalie Nichols

The Brain Maps figures provided in this section are tailored specifically for this book, encompassing regions regularly mentioned by the authors. They are therefore not intended to be comprehensive. Figures and their captions in the Brain Maps section are based on Kandel, Schwartz, Jessell, Siegelbaum, and Hudspeth (2012); Mai, Majtanik, and Paxinos (2015); and Splittgerber (2019).

Figure 1

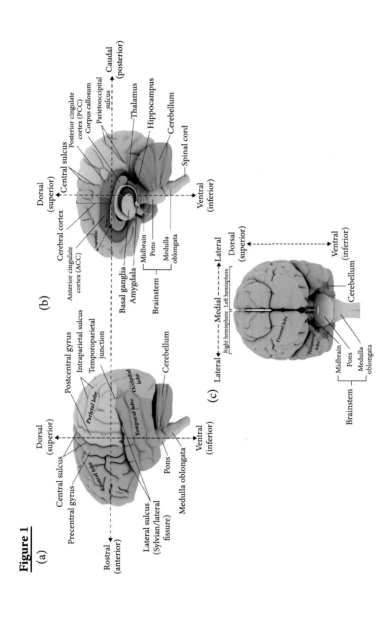

(a)

Rostral (anterior)
Dorsal (superior)
Ventral (inferior)

Central sulcus
Precentral gyrus
Postcentral gyrus
Intraparietal sulcus
Temporoparietal junction
Parietal lobe
Occipital lobe
Temporal lobe
Frontal lobe
Lateral sulcus (Sylvian/lateral fissure)
Pons
Medulla oblongata
Cerebellum

(b)

Dorsal (superior)
Ventral (inferior)
Caudal (posterior)

Central sulcus
Posterior cingulate cortex (PCC)
Corpus callosum
Parietooccipital sulcus
Thalamus
Hippocampus
Cerebellum
Spinal cord
Cerebral cortex
Anterior cingulate cortex (ACC)
Basal ganglia
Amygdala
Midbrain
Pons
Medulla oblongata
Brainstem

(c)

Lateral
Medial
Lateral
Dorsal (superior)
Ventral (inferior)
Right hemisphere
Left hemisphere
Frontal lobe
Temporal lobe
Cerebellum
Midbrain
Pons
Medulla oblongata
Brainstem

Figure 1 Three views of the brain: (a) lateral view from the left, (b) lateral view from the left with the left cortical hemisphere removed to reveal deep, subcortical structures, and (c) frontal view. The CNS is made up of the brain and *spinal cord* (1b). The brain includes the *brainstem* (1b; comprised of the *medulla oblongata* [1a, 1b, and 1c], *pons* [1a, 1b, and 1c], and *midbrain* [1b and 1c]), *cerebellum* (1a, 1b, and 1c), and cerebrum. The cerebrum is made up of the *left* and *right hemispheres* (1c). Each hemisphere, in turn, is made up of *cerebral cortex* (1b; also "cortex" —the folded outer layer of the brain made up of gyri and sulci) and deep, subcortical structures (labeled here are the *basal ganglia* [1b], *hippocampus* [1b], and *amygdaloid nuclei* or *amygdala* [1b]). The cerebral cortex is further divided into four lobes: the *frontal lobe* (pink in 1a, 1b, and 1c), *parietal lobe* (mustard yellow in 1a and 1b), *occipital lobe* (light blue in 1a and 1b), and *temporal lobe* (olive green in 1a, 1b, and 1c). Prominent sulci include the *central sulcus* (1a and 1b; separating the parietal and frontal lobes), *lateral sulcus* (1a and 1b; also termed the "lateral fissure" or "Sylvian fissure"; separating the temporal and frontal lobes), *intraparietal sulcus* (1a), and *parietooccipital sulcus* (1b; separating the occipital and parietal lobes). Prominent gyri include the *precentral gyrus* (1a; portion of the frontal lobe immediately anterior to the central sulcus) and *postcentral gyrus* (1a; portion of the parietal lobe immediately posterior to the central sulcus). The dashed double arrows indicate anatomical directions in the brain, including *rostral/anterior* and *caudal/posterior* (1a and 1b), *dorsal (superior)* and *ventral (inferior)* (1a, 1b, and 1c), *superficial* and *deep* (1b), along with *medial* and *lateral* (1c).

Figure 2

(a)

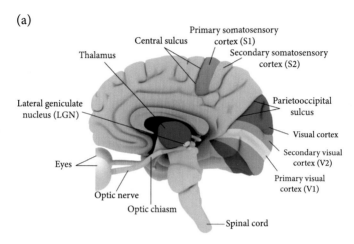

(b)

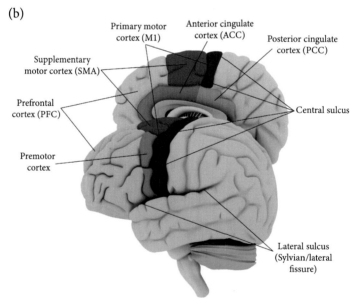

Figure 2 Some of the brain regions implicated in sensation, perception, and movement. (a) Some regions of the central nervous system (CNS) implicated in vision and somatosensation. (b) Some regions of the cerebral cortex implicated in voluntary action. Broadly speaking, visual information flows from the *eyes* (2a), through the *optic nerves* and *optic chiasm* (2a), to the *lateral geniculate nuclei* (LGN; 2a) of the *thalamus* (2a) in both hemispheres, and then to *primary visual cortex* (V1; 2a), *secondary visual cortex* (V2; 2a), other *visual cortices* (2a) in the occipital lobe (1a and 1b), and then to other parts of the brain. Body-related sensations and perceptions (e.g., the relative positions of body parts, touch, heat, pain, and tactile representation) involve afferent or ascending input from sensory receptors in the body to the spinal cord, medulla, through the *thalamus* (2a) and to somatosensory cortex. Somatosensory cortex is immediately posterior to the *central sulcus* (2a) and comprises the *primary* and *secondary somatosensory cortex* (S1 and S2, respectively; 2a). Immediately anterior to the central sulcus lies the motor cortex, which can be divided into three subregions. Roughly speaking, the *primary motor cortex* (M1; 2b) is the cortical area that controls the execution of movement most directly, through subcortical regions and the spinal cord. The *premotor cortex* (2b) is involved in several aspects of motor preparation, control, and its sensory guidance. The *supplementary motor area* (SMA; 2b) appears to be involved in self-generated action, motor sequences, bimanual coordination, and more. The *prefrontal cortex* (PFC; 2b) and cingulate cortex, especially the *anterior cingulate cortex* (ACC; 2b), are also implicated in motor control.

Figure 3

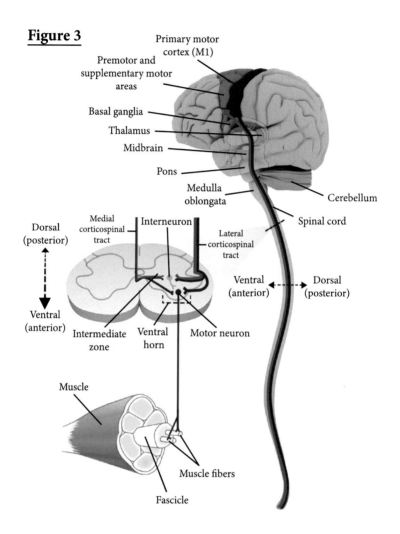

Primary motor cortex (M1)

Premotor and supplementary motor areas

Basal ganglia

Thalamus

Midbrain

Pons

Medulla oblongata

Cerebellum

Spinal cord

Dorsal (posterior)

Ventral (anterior)

Medial corticospinal tract

Interneuron

Lateral corticospinal tract

Ventral (anterior)

Dorsal (posterior)

Intermediate zone

Ventral horn

Motor neuron

Muscle

Muscle fibers

Fascicle

Figure 3 The motor system and the motor pathway. Principal components of the motor system include the *primary motor cortex, premotor and supplementary motor areas, basal ganglia, thalamus*, brainstem (*midbrain, pons, and medulla*), *cerebellum*, and *spinal cord*. The corticospinal tract is a descending pathway made up of nerve fibers (dendrites and axons of neurons; see Figure 4) that originate in the cerebral cortex (1b) and terminate in the spinal cord. Most corticospinal fibers cross the midline in the medulla and descend the spinal cord on the contralateral side of the body from where they originated (*lateral corticospinal tract*), but a small portion do not cross until they reach their destination in the spinal cord (*medial corticospinal tract*). Fibers originating in the primary motor cortex (M1) make up a considerable share of the corticospinal tract. Many corticospinal fibers that originate in the primary motor cortex (M1; 2b), along with fibers from premotor and supplementary motor area (SMA; 2b), and portions of the cingulate cortex (2b), terminate at *interneurons* in the *intermediate zone* of the spinal cord. Interneurons connect to and affect the activity of spinal *motor neurons*. Such spinal motor neurons are located in the *ventral horn* of the spinal cord and synapse onto *muscle fibers* to elicit movement. M1 is the only known motor area with neurons that connect directly to spinal motor neurons. The output of motor areas in the cortex is influenced by subcortical brain structures related to the motor system, including the cerebellum, basal ganglia, and thalamus. See Chapters 16 and 20 for more on the initiation of voluntary movement and the motor pathway. (Note that, for the spinal cord and body, "ventral" and "dorsal" refer to front and back, respectively; in contrast, for the brain and head, ventral and dorsal directions refer to up and down, respectively (1a, ab, 1c).

Figure 4

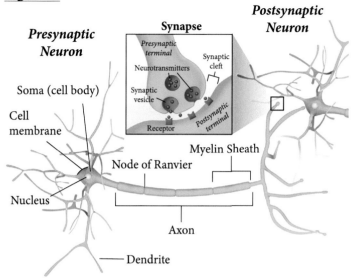

Figure 4 The structure and activity of a neuron. Neurons—broadly speaking, nerve cells structured for rapid unidirectional electrochemical signaling at a range of distances—are the basic units of the central nervous system (CNS). A neuron is generally composed of three parts: *dendrites*, a *cell body* (or soma), and an *axon*. Neurons receive input at their dendrites, while the soma processes and integrates this input. The movement of ions through ion channels in the *cell membrane* (a semi-permeable layer that forms the external boundary of the cell) is the key to all electrical events in neurons. Ion channels open or close in response to changes in or around the neuron (e.g., due to the presence of *neurotransmitters*, physical stretching of the membrane, or changes in the neuron's voltage), permitting or preventing the flow of ions into the neuron. When a neuron is sufficiently stimulated (i.e., enough ions have entered the neuron and the neuron's membrane voltage reaches its threshold), a wave of ion channels opens and ions flood into the cell. This causes the neuron to spike or fire an action potential (the primary form of communication between neurons) along its *axon* (a long structure that conducts electrical signals; it originates in the soma and is insulated by *myelin sheath*, which is broken into segments by *Nodes of Ranvier*). A bundle of axons from different cells traveling together constitutes a nerve. Neurons connect to each other via synapses. A *presynaptic neuron's* action potential affects the *postsynaptic neuron* by prompting a release of neurotransmitters from *vesicles* in the *presynaptic terminal* into the *synaptic cleft* to be collected by *receptors* on the *postsynaptic dendrite*. Neurons can be excitatory or inhibitory, meaning that when the presynaptic neuron fires an action potential, the postsynaptic neuron is more or less likely to fire its own action potential, respectively. A single neuron can also affect the activity of a population of other neurons by way of neuromodulators (a type of neurotransmitter), which alter cellular or synaptic properties of neurons and thus signal transmission.

epiphenomenalism – The claim that all mental events and states are epiphenomenal.

epistemic – Of or relating to knowledge or justification.

event-related potential (ERP) – An electrical potential (voltage) that is typically measured in relation to a stimulus or to a response (e.g., the readiness potential is an ERP measured in relation to movement onset). See *brain activity* and *readiness potential*.

excitation (neural) – A neuron or group of neurons is excited when there is a high chance that they will fire above a normal base rate. See *firing, action potential*, and Figure 4 in the Brain Maps section. Contrast *inhibition*.

experiment – An experiment is a procedure or series of steps adopted to test a hypothesis or to explore an issue through observation. Experiments often consist of a number of trials in which each participant or subject reacts to a stimulus or performs a task. Experiments are often contrasted with mere surveys on the grounds that experiments require some intervention, manipulation, or change in relevant variables. The experimental conditions are then the sets of trials when the intervention is or is not present, and the data is measurements related to different trials and conditions. For example, if the task is to push a button when but only when the circle on a screen is blue, a trial occurs when any circle appears; one condition is when the circle is blue, and the other condition is when the circle is a different color. A neuroscientist might measure brain activity just before button presses in order to test the hypothesis that we process the color blue differently than other colors. See *brain activity, neuroimaging, task*, and *behavioral*.

falsifiable – See Chapter 17.

firing (neuronal) – The release of an action potential by a neuron. See *action potential* and Figure 4 in the Brain Maps section.

first cause – A cause that is not itself caused by anything else. On some views, God and humans (or human choices) are claimed to be first causes. See *causation* and Chapters 4 and 16.

fMRI – See *functional magnetic-resonance imaging*.

free/freedom – See Chapter 4.

free-choice task – A task in which participants are instructed to choose whether, when, or what (e.g., which hand) to move. See *task* and Chapter 17.

free will – See Chapters 4 and 5.

free-will debate – A debate about the nature or existence of free will or about whether free will is compatible with certain claims (especially determinism) or necessary for other things (such as moral responsibility).

free-will task – See *free-choice task*.

functional magnetic-resonance imaging (fMRI) – Blood oxygen level dependent functional magnetic-resonance imaging (BOLD fMRI or just fMRI) is a non-invasive neuroimaging technique that records changes in blood oxygenation and blood flow that take place in response to neural activity. fMRI typically has a spatial resolution of about 1–2 mm^3. Its temporal resolution is limited because the BOLD response peaks 5–6 seconds after the onset of its corresponding neural activation. See *activity, neuroimaging, EEG*, and Chapter 27.

gyrus (plural gyri) – The crest or ridge between two deep clefts of a fold in the cerebral cortex (e.g., pre-central gyrus and post-central gyrus). Contrast *sulcus*. See Figure 1a and 1b in the Brain Maps section.

hierarchy/hierarchical – A hierarchical system is one organized into lower and higher levels. In neuroscience, this refers to a network of cortical areas or subcortical nuclei that are organized into levels of processing of information, each responsible for specific functions. See *system, network*, **and Chapter 20.**

homunculus – A homunculus is sometimes supposed to be a little person inside a brain module that controls that module. Postulating a homunculus in this sense to explain an observation is widely seen as fallacious, because it creates a new need to explain how the little person operates. In a different sense, the somatotopic mapping of body parts onto the parts of the motor and sensory strip that control correlating body parts is sometimes also called a homunculus, because it is depicted by a picture of a little body. See *somatotopic*.

hypothesis – A predicted result. See *experiment*.

incompatibilism – The claim that determinism is not compatible with free will and/or moral responsibility. See *libertarianism* and Chapters 4, 6, 7, 9, and 14.

inferior – See Figures 1a, 1b, and 1c in the Brain Maps section.

inhibition (neural) – To inhibit a neuron or group of neurons is to reduce the chance that it or they will fire. See *firing, action potential*, and Figure 4 in the Brain Maps section. Contrast *excitation*.

instructed-movement task – A task in which neither whether, when, nor what to move is up to the subject. Contrast *free-will task* and *spontaneous-movement task*. See *task*.

intention(s) – See Chapters 1, 21, 22, and 28.

invasive/non-invasive – Invasive neuroimaging or brain recording techniques involve surgically implanting a device inside or on the surface of the brain (e.g., single-unit recordings, ECoG). Non-invasive neuroimaging or brain recording techniques do not require surgical implantation (e.g., EEG, MEG, fMRI). See Chapter 27.

involuntary – See *voluntary* and Chapter 3.

ion channel – See Chapter 29 and Figure 4 in the Brain Maps section.

ipsilateral – Relates to the same side. For example, brain activity in the right frontal lobe is ipsilateral to brain activity in the right parietal lobe, but is not ipsilateral to any activity in the left hemisphere. See Figure 1c in the Brain Maps section. Contrast *contralateral* and *bilateral*.

lateral – See Figure 1c in the Brain Maps section.

lesion – In neuroscience, lesion refers to damage to a part of the brain (including damage due to stroke, hemorrhage, trauma, disease, or surgery, for example).

libertarian – Libertarianism is the thesis that some agents, actions, and wills or decisions are free even though freedom of action and will are incompatible with determinism or with some sources of action like manipulation. Libertarianism is thus a form of incompatibilism that denies determinism of the relevant kind. Libertarian freedom is the kind of freedom that is claimed by libertarians. Libertarianism about free will should not be confused with libertarianism as a political view. See *incompatibilism* and Chapters 9 and 10.

Libet – Benjamin Libet (1916–2007) was a pioneering neuroscientist of volition and consciousness who is well-known for "Libet experiments" that informed the neuroscience of volition. See Chapters 19 and 28.

Libet-like task – See *spontaneous-movement task*, along with Chapters 19 and 28.

Libet's W-time – See Chapter 28.

liking – Disposed or expecting to get pleasure or some positive feeling from obtaining and achieving what one likes. Some agents are sometimes said to want something that they do not like. See *wanting* and Chapter 23.

limbic system – A large collection of cortical and subcortical regions mainly around the thalamus, supporting functions like emotion, behavior, motivation, long-term memory, and olfaction. See Figure 1b in the Brain Maps section.

locked-in syndrome (LIS) – A condition by which a patient retains awareness but suffers complete paralysis of all voluntary muscles except some residual eye movement. If eye movement too is absent, then it is complete or total LIS.

magnetoephalography (MEG) – A non-invasive neuroimaging technique that records magnetic fields produced by electric currents in the brain using highly sensitive magnetometers on the scalp. Its temporal resolution is similar to EEG, though its spatial resolution is sometimes claimed to be higher than EEG because magnetic fields undergo less distortion between the brain and scalp. See *neuroimaging* and Chapter 27.

mechanism – A system whose parts and processes causally interact to bring about one or more effects. In the context of neuroscience, a neural mechanism typically refers to a mechanism instantiated by a neural network. See *system* and *causation*.

medial – See Figure 1c in the Brain Maps section.

MEG – See *magnetoephalography*.

mental representation – See *representation*.

mental state – A state of the mind at a particular time (such as my anger at noon today) or a type or kind of such states (such as anger). Mental states are sometimes contrasted with mental events that are changes in the mind over time, such as making a decision.

mind-body problem – The issue of how minds (or mental states and events) are related to bodies (or brains or physical states and events). The main proposed solutions to this problem include dualism, emergentism, and reductionism. See *dualism, emergentism, reductionism*, and Chapter 25.

model – A purposeful simplification and abstraction of a certain group or kind of observed phenomena. Models in the exact sciences often rely on mathematical formulations (mathematical models), in contrast to models that rely on verbal descriptions (verbal models). See Chapter 30.

motoneuron – See *motor neuron*.

motor command – See Chapter 16.

motor control – The regulation, control, or coordination of voluntary movement via the nervous system. See *control*.

motor neuron – A neuron located in the motor cortex, brainstem, or the spinal cord that projects to the spinal cord or outside of the spinal cord and either directly or indirectly influences muscles and glands. Also called a motoneuron. See Chapters 16 and 20, along with Figures 3 and 4 in the Brain Maps section.

motor system – The group of brain systems that regulate, control, or coordinate voluntary bodily movements. See Figures 2b and 3 in the Brain Maps section.

necessary condition – One thing is a necessary condition of another when the latter cannot occur without the former. Contrast *sufficient condition*.

network – A biological neural network is a collection of neurons that carries out a certain function and in which at least some of the neurons connect to others via synapses. These have inspired artificial neural networks, which are non-biological collections of at least partially connected units that transmit information to each other. See Figure 4 in the Brain Maps section.

neural correlate – Brain activity that corresponds in a systematic way to a certain experience or function and is necessary for that experience or function. In particular, the neural correlates of consciousness are the minimal neuronal mechanisms that are jointly sufficient for any one specific conscious percept. See *brain activity* and Chapters 16, 23, and 25.

neural representation – See *representation*.

neuroimaging – Methods or techniques for recording signals related to brain activity. See *brain activity, ECoG, EEG, fMRI, PET, single-unit recording,* and *SPECT*. See also Chapter 27.

neuromodulator – See Figure 4 in the Brain Maps section.

neuron – See Figure 4 in the Brain Maps section.

neurotransmitter – See Figure 4 in the Brain Maps section.

noise – In neuroscience, noise (in contrast to "signal") refers to random, intrinsic fluctuations in neural activity that are not related to the task being carried out by the experimental participant. See *random, activity, experiment*, and Chapter 29.

no-report paradigm – A paradigm that tries to infer the content of subjects' conscious experience without relying on their self-reports of those experiences. Contrast *subjective measures/report*. See Chapter 24.

operational definition – See Chapters 17 and 24.

order/disorder – See Chapter 29.

PET – See *positron emission tomography*.

phenomenal consciousness – An entity has phenomenal consciousness to the extent that there is something it is like to be that entity. It is usually held that humans have phenomenal consciousness, but rocks do not, because humans have subjective or essentially first-personal experiences, but rocks do not. Contrast *access consciousness*. See Chapter 12.

phenomenology – Phenomenology as a discipline studies structures of consciousness from the first-person point of view. The phenomenology of a particular experience or a kind of experience is what it is like to have that experience or that kind of experience.

physicalism – The view that everything that exists is physical in some way. Some physicalists allow that physical things, including brains, can have mental properties. See *dualism, emergentism,* and *reductionism*. See Chapters 25 and 26.

picking – See *arbitrary decisions*.

positron emission tomography (PET) – An imaging technique that uses radioactive materials (or tracers) to visualize metabolic processes in the body and in particular in the brain. See Chapter 27.

posterior – See Figures 1a and 1b in the Brain Maps section.

praxis – See Chapter 18.

proximal – See *decide/decision* and Chapters 21, 22, 25 and 28.

quale/qualia – Quale is the singular of qualia. Qualia are subjective aspects of our mental lives, including what it is like to be in a mental state (such as to see red or to make a decision). Qualia are thus signs of phenomenal consciousness. See *phenomenal consciousness* and Chapter 25.

random – Arbitrary; lacking any pattern, method, or systematic qualities.

rational/rationality – Rationality is the ability of people to be rational. Agents are rational to the extent that their actions are rational. Actions are rational to the extent that they are based on reasons (or at least not against reason) or to the extent that they result from intentions, desires, and beliefs that are minimally informed, consistent, and/or proper. Philosophers and others disagree about which specific conditions must be met for a person or act to be rational. See Chapter 15.

reaction-time task – A task in which participants are instructed to move as quickly as possible after the onset of some cue. See *task*.

readiness potential (also RP, Bereitschaftspotential, or BP) – An event-related potential (ERP) characterized by a slow negative shift that is most strongly measured at frontal and central electrode sites over motor areas in the brain leading up to voluntary action. Some scientists distinguish between different parts of the RP (early RP/BP or BP1 and late RP/BP, BP2, or NS'). The equivalent measure for MEG is termed "readiness field." See *event-related potential, EEG, MEG*, and Chapters 19, 22, 27, and 28.

real time – Refers to measuring or decoding brain activity via neuroimaging techniques on the fly. This is done, for example, in order to decode a participant's present mental state or impending behavior while the experiment is in session. See *decoding, brain activity, neuroimaging*, and *experiment*.

reason-responsiveness – See Chapters 4, 15, and 23.

reasons/reasoning – See Chapter 15.

reductionism – Reductionism as a view about the mind-body problem holds that minds along with mental states and events are not identical with or reducible to neural or other physical states or events. Token-token reductionism is the view that individual or particular instances of mental states and events are identical with and reducible to individual or particular instances of physical states and events. Type-type reductionism is the view that general mental properties are identical with and reducible to general physical properties. Some philosophers hold token-token reductionism but deny type-type reductionism. See *dualism* and *emergentism*. See Chapter 26.

reinforcement learning – An adaptive process by which an animal utilizes its experience to improve future choices. It is also an area in machine learning that is concerned with adapting software agents to act in an environment to maximize some reward. See *reward*.

representation – A representation of something is anything that is about or stands for that thing, such as a painting of Mount Everest or the words "Mount Everest." A *mental* representation is then a mental state or event that is about or stands for something, usually in the external world, such as a perception or memory of Mount Everest. A *neural* representation is a brain state or pattern of brain activity that reflects or stands for something, such as an object or a concept.

responsible/responsibility – See Chapter 14.

reward – An event that reinforces a behavior when it is contingent upon that behavior.

rostral – See Figures 1a and 1b in the Brain Maps section.

RP – See *readiness potential.*

self-control – Control over one's own actions or thoughts. See *control.*

self-initiated movement task – See *spontaneous-movement task.*

semi-compatibilism – Strong semi-compatibilists claim that determinism (and the kind of inability to choose and act otherwise that determinism implies) is compatible with moral responsibility but is incompatible with freedom of action or will. A weaker version of semi-compatibilism takes no stand on freedom or its compatibility with determinism and asserts only that determinism and moral responsibility are compatible. See Chapter 14.

sense of agency – The experience of having caused or of being able to perform an action. See Chapter 25.

single-photon emission computed tomography (SPECT) – An imaging technique similar to PET, though it depends on directly detecting gamma rays emitted from a radio tracer rather than those emitted from positron-electrode annihilations. Its spatial resolution is lower than PET. See *positron emission tomography (PET), brain activity,* and Chapter 27.

single-unit – Single-unit recording refers to measuring the activity of single neurons (namely, the spiking or firing of action potentials) with invasive electrodes. See *firing, action potential,* and Figure 4 in the Brain Maps section.

somatosensory system – Part of the sensory nervous system that responds to changes at or inside the body (e.g., body position, touch, temperature). See Figure 2a in the Brain Maps section.

somatotopic – Somatotopy is the mapping of the body onto specific locations in the brain that receive input from or control those parts of the body. Such mappings exist in somatosensory cortex for sensory inputs and in motor cortex for motor processing and muscle control.

spatial resolution – A measure of the smallest object that can be resolved by a sensor. In other words, it refers to how fine-grained a digital image is. Higher spatial resolution means a more fine-grained image. Contrast *temporal resolution.* See *fMRI, EEG,* and Chapter 21.

SPECT – See *single-photon emission computed tomography*.

speeded reaction-time task – See *reaction-time task*.

spontaneous – See Chapters 2 and 18.

spontaneous-movement task – A task in which the timing of the movement is left up to the subject. Contrast *instructed-movement task* and *reaction-time task*. See *task* and Chapter 18.

stimuli – See *experiment*.

stochastic – A stochastic process is one that is not determined or controlled by deterministic causes, typically because it encompasses some random fluctuations. As a result, the process and its outcome cannot be tracked or predicted with absolute precision. A stochastic process can still sometimes be tracked by probability distributions, so its pattern can be analyzed statistically (e.g., its mean and standard deviation might be calculable). Contrast *determinism*. See Chapter 29.

subjective – See *quale/qualia* and Chapter 25.

subjective report (or self-report) – See Chapters 9 and 24.

sufficient condition – One thing is a sufficient condition of another when the former cannot occur without the latter. Contrast *necessary condition*.

sulcus (plural sulci) – Also called "fissure(s)." The groove, furrow, cleft, or depressed portion between folds on the cerebral cortex (e.g., central sulcus or lateral sulcus). Contrast *gyrus*. See Figure 1a and 1b in the Brain Maps section.

superficial – See Figure 1b in the Brain Maps section.

superior – See Figures 1a, 1b, and 1c in the Brain Maps section.

synapse/synaptic – See Figure 4 in the Brain Maps section.

system – An organized set of components that interact in various ways to perform a somewhat unified function (e.g., the visual system or motor system in the brain). See *network*.

task – What a participant is instructed to do in an experiment. See *experiment* and *behavioral*.

temporal resolution – Measuring equipment often measures a signal at constant intervals. Those constant intervals are the temporal resolution of that measurement. The higher the temporal resolution (i.e., the closer together in time that the signal is measured), the more the measurement will capture fine-grained changes in the signal. Contrast *spatial resolution*. See *fMRI*, *EEG*, and Chapter 21.

threshold – In neuroscience, this typically refers to some minimum activation, energy, voltage, and so on above which something of interest happens. For example, when a neuron's membrane potential exceeds a certain voltage—its activation threshold—the neuron fires an action potential. Some models of decision-making also postulate a threshold of neural activity at which a decision is made. See *model*, Chapter 30, and Figure 4 in the Brain Maps section.

top-down – Moving from a higher level to a lower level. Information processing is described as top-down when it starts with abstract or conceptual information about general kinds of events or experiences and results in concrete information about individual events or experiences. Contrast *bottom-up*.

transcranial magnetic stimulation (TMS) – A non-invasive brain-stimulation or perturbation method that uses focused magnetic fields to stimulate a specific area of the brain. TMS of primary motor cortex (see Figures 2b and 3 in the Brain Maps section) may result in involuntary movements. It is more common for TMS to inhibit rather than excite brain activity. See *invasive/non-invasive* and Chapter 27.

ventral – See Figures 1a, 1b, and 1c in the Brain Maps section.

veto – In the context of neuroscience, it is the capacity to halt or cancel an impending movement. In relation to the Libet experiment in particular, it refers to the idea that the only role for consciousness in action formation is to potentially stop, or veto, unconsciously initiated behavior. See *Libet* and Chapters 22 and 25.

visual system – Comprises the eyes and vision-processing parts of the central nervous system (CNS; optic nerve, lateral geniculate nucleus [LGN], and visual cortex [V1–V5 or Brodmann's areas 17–19]). See Figure 2a in the Brain Maps section.

volition – The words "volition" and "will" are often used almost interchangeably. See Chapters 2, 17, and 26.

volition(al) task – See *free-choice task*.

voluntary – An act is voluntary to the extent that it results from volition or will. See *volition*, *will*, and Chapter 3.

wanting – A motivational state related to desire. Some agents are sometimes said to want something that they do not like. Contrast *liking*. See *desire* and Chapter 23.

weakness of will – See *akrasia*.

will – Will is the capacity for choice or decision or, more specifically, an exercise of that capacity. See Chapters 2, 17, and 26.

Annotated Bibliography

Deniz Arıtürk, Amber Hopkins, and Claire Simmons

Ach, N. (1910/2006). *On Volition* (T. Herz, Trans.). University of Konstanz.
(Lecture summarizing the ideas of German psychologist Ach, whose experiments on learned associations and task goals are considered by some to be the predecessor of conflict paradigms like the Stroop task; suitable for readers with moderate background in psychology)

Adams, R. A., Shipp, S., & Friston, K. J. (2013). Predictions not commands: Active inference in the motor system. *Brain Structure and Function, 218*(3), 611–643.
(Describes the functional anatomy of the motor system from an active inference perspective, which argues that descending pathways carry predictions rather than commands; suitable for readers with strong background in neuroscience)

Ajina, S., & Bridge, H. (2017). Blindsight and unconscious vision: What they teach us about the human visual system. *The Neuroscientist, 23*(5), 529–541. doi:10.1177/1073858416673817
(Reviews the residual abilities and neural activity that have been described in blindsight, wherein some patients retain the ability to detect visual information within the blind region of the visual cortex without being aware of that ability, and discusses the implications of these findings for understanding the intact visual system; suitable for readers for minimal background in neuroscience)

Allman, M. J., Teki, S., Griffiths, T. D., & Meck, W. H. (2014). Properties of the internal clock: First- and second-order principles of subjective time. *Annual Review of Psychology, 65*, 743–771.
(Discusses subjective time passing, or the internal clock, reviewing relevant behavioral and neurobiological findings; suitable for readers with minimal background in neuroscience)

Anscombe, G. E. M. (1957). *XIV.—Intention.* Paper presented at the Proceedings of the Aristotelian Society.
(Distinguishes actions that are intentional from those that are not by appealing to "why" questions and the distinction between reasons and causes; suitable for readers with moderate background in philosophy)

Anscombe, G. E. M. (2000). *Intention.* Harvard University Press.

(Explains what intentions are and how they differ from other mental states like beliefs or predictions about the future in important ways, including in their "direction of fit"; suitable for readers with moderate background in philosophy)

Århem, P., Blomberg, C., & Liljenström, H. (2000) . Disorder versus order in brain function: Essays in theoretical neurobiology. *Progress in Neural Processing, 12.* World Scientific Publ. Co.

(Compilation of essays on order and disorder in the brain, covering many topics, from noise and chaos, neural computations, fluctuations, and randomness, to perception, consciousness, schemata, and language; suitable for readers with strong background in neuroscience)

Århem, P., Braun, H. A., Huber, M., & Liljenström, H. (2005). Non-linear state transitions in neural systems: From ion channels to networks. In H. Liljenström & U. Svedin (Eds.), *Micro—Meso—Macro: Addressing Complex Systems Couplings* (pp. 37–72). World Scientific Publ. Co.

(Using experimental and computational methods, investigates ion channel kinetics and how neural systems amplify weak signals to modulate the system at a larger scale; suitable for readers with strong background in neuroscience)

Århem, P., & Liljenström, H. (2001). Fluctuations in neural systems: From subcellular to network levels. In F. Moss & S. Gielen (Eds.), *Handbook of Biological Physics* (Vol. 4, pp. 83–129). North-Holland.

(Reviews the underlying mechanisms and the possible functional role of fluctuations in the nervous systems, which provide a way to obtain information on low-level processes by studying activity at a higher level; suitable for readers with moderate background in neuroscience)

Aristotle. (1999). *Nicomachean Ethics* (T. Irwin, Ed. Second ed.). Oxford University Press.

(Among many other topics in ethics, divides actions into three categories—voluntary, involuntary, and non-voluntary—and explores the ethical implications of each; suitable for readers with minimal background in philosophy)

Bai, O., Rathi, V., Lin, P., Huang, D., Battapady, H., Fei, D.-Y., . . . Hallett, M. (2011). Prediction of human voluntary movement before it occurs. *Clinical Neurophysiology, 122*(2), 364–372.

(Uses EEG and machine learning to predict impending movement in real-time while human subjects perform self-paced movements; suitable for readers with moderate background in neuroscience)

Banks, W. P., & Isham, E. A. (2009). We infer rather than perceive the moment we decided to act. *Psychological Science, 20*(1), 17–21.

(Uses a deceptive-feedback experiment to demonstrate that W, the reported time of decision used in paradigms like the Libet experiment, moves forward in time linearly with the delay in feedback, and concludes that participants' report of their decision time is largely inferred from the apparent time of response; suitable for readers with minimal background in neuroscience)

Bartra, O., McGuire, J. T., & Kable, J. W. (2013). The valuation system: A coordinate-based meta-analysis of BOLD fMRI experiments examining neural correlates of subjective value. *Neuroimage, 76,* 412–427. doi:https://doi.org/10.1016/j.neuroimage.2013.02.063

(Meta-analysis of fMRI studies concluding that the ventromedial prefrontal cortex and anterior ventral striatum constitute a valuation system in decision making; suitable for readers with minimal background in neuroscience)

Baumeister, R. F., & Vohs, K. D. (2003). Willpower, choice, and self-control. In G. Loewenstein, D. Read, & R. Baumeister (Eds.), *Time and Decision: Economic and Psychological Perspectives on Intertemporal Choice* (pp. 201–216). Russell Sage Foundation.

(Discusses research on self-control and willpower and their relation to quality of life; suitable for readers with no background in neuroscience)

Bayne, T. (2008). The phenomenology of agency. *Philosophy Compass, 3,* 182–202.

(Presents an overview of recent discussions on the phenomenology of agency; suitable for readers with minimal background in philosophy)

Bayne. T. (2017). Free will and the phenomenology of agency. In K. Timpe, M. Griffith, & N. Levy (Eds.), *The Routledge Companion to Free Will* (pp. 633–644). Routledge.

(Explores the genetic question of free will (Why do people believe that they have free will?) and the epistemic question of free will (Is the belief in free will well-founded?) and finds them to be intimately related; suitable for readers with moderate background in philosophy)

Beck, F., & Eccles, J. C. (1992). Quantum aspects of brain activity and the role of consciousness. *Proceedings of the National Academy of Sciences, 89*(23), 11357–11361. doi:10.1073/pnas.89.23.11357

(Presents a quantum mechanical description of brain structure and activity, argues conscious states alter brain activity and cause action; suitable for readers with moderate background in neuroscience)

Blanchette, I. (2014). What is the role of the ventromedial prefrontal cortex in emotional influences on reason? In L. Blanchette (Ed.), *Emotion and Reasoning* (pp. 154–173). Psychology Press.

(Book chapter on the involvement of the vmPFC in both emotional and moral reasoning; suitable for readers with minimal background in neuroscience)

Block, N. (2002). Concepts of consciousness. In D. J. Chalmers (Ed.), *Philosophy of Mind: Classical and Contemporary Readings* (pp. 206–218). Oxford University Press.

(Defines various concepts of consciousness, including phenomenal con-sciousness and access consciousness; suitable for readers with moderate background in philosophy)

Block, N. (2007). Consciousness, accessibility, and the mesh between psy-chology and neuroscience. *Behavioral and Brain Sciences, 30*(5–6), 481–548.

(Relies on empirical data to propose a way to disentangle the neural basis of phenomenal consciousness from the neural machinery of cogni-tive access that underlies reports of phenomenal consciousness; suitable for readers with strong background in philosophy and moderate background in neuroscience)

Bodranghien, F., Bastian, A., Casali, C., Hallett, M., Louis, E. D., Manto, M., . . . Serrao, M. (2016). Consensus paper: Revisiting the symptoms and signs of cerebellar syndrome. *The Cerebellum, 15*(3), 369–391.

(Reviews the effects of cerebellar dysfunction on cognition, motor op-erations, proprioception, and affective regulation, among other processes; suitable for readers with minimal background in neuroscience)

Bratman, M. (1987). *Intention, Plans, and Practical Reason.* Harvard University Press.

(Maps the ways in which intentions differ from other mental states that share their direction of fit in light of its effects on behavior and reasoning; suitable for readers with moderate background in philosophy)

Bréchet, L., Brunet, D., Birot, G., Gruetter, R., Michel, C. M., & Jorge, J. (2019). Capturing the spatiotemporal dynamics of self-generated, task-initiated thoughts with EEG and fMRI. *Neuroimage, 194*, 82–92.

(Using fMRI and EEG data, reports a correlation of brief, identifiable neuronal activity with conscious memory phenomena; suitable for readers with moderate background in neuroscience)

Brown, R., Lau, H., & LeDoux, J. E. (2019). Understanding the higher-order approach to consciousness. *Trends in Cognitive Sciences, 23*(9), 754–768. doi:https://doi.org/10.1016/j.tics.2019.06.009

(Presents a clear overview of and clarifies misunderstandings about the higher-order approach, one of the main theories of consciousness that allows for substantial unconscious processing prior to conscious pro-cessing; suitable for readers with minimal background in philosophy and moderate background in neuroscience)

Caminiti, R., Borra, E., Visco-Comandini, F., Battaglia-Mayer, A., Averbeck, B. B., & Luppino, G. (2017). Computational architecture of the parieto-frontal network underlying cognitive-motor control in monkeys. *eneuro, 4*(1), ENEURO.0306-0316.2017. doi:10.1523/eneuro.0306-16.2017

(Uses hierarchical cluster analysis on macaque anatomic connectivity data to investigate connections between parietal and frontal lobes and to pinpoint processing streams involved in motor control and coordination; suitable for readers with moderate background in neuroscience)

Carroll, J. W. (2016). Laws of nature. In E. N. Zalta (Ed.), *The Stanford Encyclopedia of Philosophy* (Fall 2016 ed.). https://plato.stanford.edu/archives/win2020/entries/laws-of-nature/
(A discussion of the two competing visions of natural laws as understood by compatibilists and incompatibilists; suitable for readers with minimal background in philosophy)

Caspar, E. A., Christensen, J. F., Cleeremans, A., & Haggard, P. (2016). Coercion changes the sense of agency in the human brain. *Current Biology*, *26*(5), 585–592.
(Presents human subjects with coercive instructions to harm co-participants, reports that coercion inflates the perceived time interval between action and outcome (an implicit measure of agency) and reduces action-outcome processing; suitable for readers with minimal background in neuroscience)

Castiello, U., Paulignan, Y., & Jeannerod, M. (1991). Temporal dissociation of motor responses and subjective awareness: A study in normal subjects. *Brain, 114* (Pt 6), 2639–2655. doi:10.1093/brain/114.6.2639
(Reports a delay between motor response to and conscious awareness of the sudden displacement of a visual object and concludes that neural activity is processed before it gives rise to conscious experience; suitable for readers with strong background in neuroscience)

Chalmers, D. J. (2015). Why isn't there more progress in philosophy? *Philosophy*, *90*(1), 3–31.
(Articulates and argues for a version of the "glass-half-empty view" that there is not as much progress in philosophy as we would like; suitable for readers with minimal background in philosophy)

Clark, A. (1991). *Microcognition: Philosophy, Cognitive Science, and Parallel Distributed Processing*. MIT Press.
(Explains and explores the biological basis of parallel distributed processing, its psychological importance, and its philosophical relevance for matters including intentionality; suitable for readers with moderate background in philosophy and neuroscience)

Correa, C. M., Noorman, S., Jiang, J., Palminteri, S., Cohen, M. X., Lebreton, M., & van Gaal, S. (2018). How the level of reward awareness changes the computational and electrophysiological signatures of reinforcement learning. *Journal of Neuroscience*, *38*(48), 10338–10348.
(Uses a computational model trained on EEG from human subjects completing a reinforcement learning task to show that reward awareness plays a role in adjusting decision-making strategies; suitable for readers with moderate background in neuroscience)

Crick, F. (1994). *The Astonishing Hypothesis: The Scientific Search for the Soul* (Vol. 37). Scribner.

(Nobel laureate Francis Crick dives into the concept of the human soul, its purpose, and the human brain; suitable for readers with minimal background in neuroscience)

Debner, J. A., & Jacoby, L. L. (1994). Unconscious perception: Attention, awareness, and control. *Journal of Experimental Psychology: Learning, Memory, and Cognition, 20*(2), 304–317. doi:10.1037/0278-7393.20.2.304

(A famous psychology experiment that employs a "process-dissociation procedure" to demonstrate that unconscious perception can influence behavior; suitable for readers with moderate background in psychology)

Dehaene, S., Changeux, J.-P., Naccache, L., Sackur, J., & Sergent, C. (2006). Conscious, preconscious, and subliminal processing: A testable taxonomy. *Trends in Cognitive Sciences, 10*(5), 204–211. doi:https://doi.org/10.1016/j.tics.2006.03.007

(On the basis of the global neuronal workspace theory of consciousness, proposes a taxonomy that distinguishes between vigilance and access to conscious report, as well as between subliminal, preconscious, and conscious processing; suitable for readers with moderate background in neuroscience)

Deiber, M.-P., Honda, M., Ibañez, V., Sadato, N., & Hallett, M. (1999). Mesial motor areas in self-initiated versus externally triggered movements examined with fMRI: Effect of movement type and rate. *Journal of Neurophysiology, 81*(6), 3065–3077.

(Using fMRI while human subjects perform self-initiated or cued movements, concludes that frontomesial motor areas are implicated in mode of movement initiation, but also movement type and rate; suitable for readers with moderate background in neuroscience)

Dennett, D. C. (1991). *Consciousness Explained.* Penguin Books.

(Provides accessible analyses of examples, empirical evidence, thought experiments, and arguments to refute dualism and propose a new model of consciousness, the Multiple Drafts model, which is based on an analogy to a computer that processes information in parallel; suitable for readers with minimal background in philosophy)

Descartes, R. (1641/1984). *Meditations on First Philosophy.* Caravan Books.

(A turning point in the history of Western thought in which Descartes, among other things, argues for a form of interactive substance dualism about mind and body; suitable for readers with moderate background in philosophy)

Desmurget, M., Reilly, K. T., Richard, N., Szathmari, A., Mottolese, C., & Sirigu, A. (2009). Movement intention after parietal cortex stimulation in humans. *Science, 324*(5928), 811–813.

(Direct electrical brain stimulation demonstrates will to move and agency over actions that do not take place and actions without will to move or even awareness of having moved; suitable for readers with moderate background in neuroscience)

Destexhe, A., & Rudolph-Lilith, M. (2012). *Neuronal Noise* (Vol. 8). Springer Science & Business Media.
(Book on noise in the brain, discussing how neuronal noise is modeled and studied; suitable for readers with moderate background in neuroscience)

Dretske, F. (1988). *Explaining Behavior: Reasons in a World of Causes.* MIT Press.
(Provides an overview of and attempts to reconcile the different scientific viewpoints on the way reasons function in the causal explanation of behavior; suitable for readers with moderate background in philosophy and minimal background in psychology)

Eagleman, D. M. (2008). Human time perception and its illusions. *Current Opinion in Neurobiology, 18*(2), 131–136.
(Reviews literature investigating temporal illusions of duration, temporal order, and simultaneity; suitable for readers with minimal background in neuroscience)

Faivre, N., Mudrik, L., Schwartz, N., & Koch, C. (2014). Multisensory integration in complete unawareness: Evidence from audiovisual congruency priming. *Psychological Science, 25*(11), 2006–2016. doi:10.1177/0956797614547916
(Presents the first empirical study suggesting that following conscious learning, unconscious processing suffices for multisensory integration across auditory and visual modalities; suitable for readers with minimal background in neuroscience)

Favilla, M., Gordon, J., Hening, W., & Ghez, C. (1990). Trajectory control in targeted force impulses. *Experimental Brain Research, 79*(3), 530–538.
(Uses the "timed response paradigm" to support a previous demonstration that subjects may perform different component processes of sensorimotor processing, such as specification of response amplitude and direction, by parallel processing channels; suitable for readers with moderate background in neuroscience)

Feyerabend, P. (1993). *Against Method.* Verso.
(Shows deficiencies of some widespread ideas about scientific progress and considers whether anarchism could replace rationalism in the theory of knowledge; suitable for readers with moderate background in philosophy)

Fischer, J. (1988). Responsiveness and moral responsibility. In F. Schoeman (Ed.), *Responsibility, Character, and the Emotions: New Essays in Moral Psychology* (pp. 81–106). Cambridge University Press.
(Introduces the "semi-compatibilist" position on free will, which claims that the abilities to choose and act otherwise are not necessary for moral responsibility; suitable for readers with minimal background in philosophy)

Fischer, J. M., & Ravizza, M. (1998). *Responsibility and Control: A Theory of Moral Responsibility.* Cambridge University Press.

(Provides a systematic compatibilist theory of moral responsibility based on "guidance control" and reasons-responsiveness; suitable for readers with moderate background in philosophy)

François-Brosseau, F. E., Martinu, K., Strafella, A. P., Petrides, M., Simard, F., & Monchi, O. (2009). Basal ganglia and frontal involvement in self-generated and externally-triggered finger movements in the dominant and non-dominant hand. *European Journal of Neuroscience, 29*(6), 1277–1286.

(Uses fMRI on human subjects completing self-initiated or externally triggered movements with their dominant or nondominant hand, concluding that the putamen is recruited for non-routine movement execution; suitable for readers with minimal background in neuroscience)

Frankfurt, H. (1969). Alternate possibilities and moral responsibility. *Journal of Philosophy, 66*(23), 829–839.

(Introduces "Frankfurt examples," which suggest that the ability to do otherwise is not necessary for moral responsibility; suitable for readers with moderate background in philosophy)

Frankfurt, H. (1988a). Freedom of the will and the concept of a person. In *The Importance of What We Care About: Philosophical Essays* (pp. 11–25). Cambridge University Press.

(A compatibilist argument that free will depends on the agent's second-order desires; suitable for readers with moderate background in philosophy)

Frankfurt, H. (1988b). The importance of what we care about. In *The Importance of What We Care About: Philosophical Essays* (pp. 80–94). Cambridge University Press.

(Argues that willed activity is free not because it is independent but because it is distinctively "owned by" the individual whose activity it is; suitable for readers with moderate background in philosophy)

Frankfurt, H. (1988c). The problem of action. In *The Importance of What We Care About: Philosophical Essays* (pp. 69–79). Cambridge University Press.

(Identifies action as purposive (voluntary) movement guided by an agent, as opposed to purposive movement guided by something other than an agent, such as reflexes or anatomical responses that are involuntary; suitable for readers with moderate background in philosophy)

Freeman, W. J. (1975). *Mass Action in the Nervous System.* Academic Press.

(Examines the neurophysiological basis of adaptive behavior through the EEG by focusing on its neural mechanisms and behavioral significance, with emphasis on observations made on the mammalian olfactory system; suitable for readers with strong background in neuroscience)

Freeman, W. J. (1996). Random activity at the microscopic neural level in cortex ("noise") sustains and is regulated by low-dimensional dynamics of macroscopic cortical activity ("chaos"). *International Journal of Neural Systems, 7*(4), 473–480.

(Describes fluctuations and spontaneous patterns of neural activity at microscopic and macroscopic neural levels of functioning and how such activity interacts; suitable for readers with strong background in neuroscience)

Freeman, W. J. (2000). *Neurodynamics: An Exploration in Mesoscopic Brain Dynamics*. Springer.

(Presents data, models, and experimental techniques comprising a mesoscopic approach to brain dynamics, which conceives, identifies, and models intervening levels between the neuron and the brain; suitable for readers with moderate background in neuroscience)

Fridman, E. A., Immisch, I., Hanakawa, T., Bohlhalter, S., Waldvogel, D., Kansaku, K., . . . Hallett, M. (2006). The role of the dorsal stream for gesture production. *Neuroimage, 29*(2), 417–428.

(Uses fMRI while human subjects make transitive vs. intransitive hand movements to conclude that the dorsal stream is recruited for mechanical knowledge and the ventral premotor and posterior parietal cortices are important to sensorimotor processing; suitable for readers with minimal background in neuroscience)

Fried, I., Haggard, P., He, B. J., & Schurger, A. (2017). Volition and action in the human brain: Processes, pathologies, and reasons. *Journal of Neuroscience, 37*(45), 10842–10847.

(Reviews literature on the generation of voluntary action and dives into three features of volition: its apparent spontaneity, its link to conscious experience, and its goal-directedness; suitable for readers with minimal background in philosophy and minimal background in neuroscience)

Fried, I., Katz, A., McCarthy, G., Sass, K. J., Williamson, P., Spencer, S. S., & Spencer, D. D. (1991). Functional organization of human supplementary motor cortex studied by electrical stimulation. *Journal of Neuroscience, 11*(11), 3656–3666.

(Using direct, electrical brain stimulation in human subjects, highlights the somatotopic organization of supplementary motor area and its role in generating an urge to move; suitable for readers with moderate background in neuroscience)

Fried, I., Mukamel, R., & Kreiman, G. (2011). Internally generated preactivation of single neurons in human medial frontal cortex predicts volition. *Neuron, 69*(3), 548–562.

(Records single neurons in human subjects completing self-initiated movements to identify neuronal activity in the supplementary motor area that is predictive of decision onset and decision awareness onset; suitable for readers with strong background in neuroscience)

Furstenberg, A., Breska, A., Sompolinsky, H., & Deouell, L. Y. (2015). Evidence of change of intention in picking situations. *Journal of Cognitive Neuroscience, 27*(11), 2133–2146.

(Provides evidence for effect of unconscious priming in free-choice task on reaction time and choice preference; suitable for readers with moderate background in neuroscience)

Gert, B., & Duggan, T. J. (1979). Free will as the ability to will. *Noûs*, *13*(2), 197–217. doi:10.2307/2214397

(A compatibilist argument that freedom requires the ability to respond to reasons; suitable for readers with moderate background in philosophy)

Goel, V., & Dolan, R. J. (2003). Explaining modulation of reasoning by belief. *Cognition*, *87*(1), B11–B22. doi:https://doi.org/10.1016/S0010-0277(02)00185-3

(Using fMRI in human subjects performing reasoning tasks, investigates the neural correlates of logic-based reasoning (lateral prefrontal cortex), belief-based reasoning, and belief bias (ventral medial prefrontal cortex); suitable for readers with minimal background in neuroscience)

Goldberg, G. (1985). Supplementary motor area structure and function: Review and hypotheses. *Behavioral and Brain Sciences*, *8*(4), 567–588. doi:10.1017/S0140525X00045167

(Synthesizes perspectives on supplementary motor area function to suggest that the SMA is implicated in intentional processes, action specification, and action elaboration; suitable for readers with minimal background in neuroscience)

Goschke, T. (2013). Volition in action. In W. Prinz, M. Beisert, & A. Herwig (Eds.), *Action Science: Foundations of an Emerging Discipline* (pp. 409–435). MIT Press.

(Addresses the question of how voluntary actions are determined and argues that intentions are not direct triggering causes of actions but rather internal constraints biasing perceptual processing and response selection over extended durations; suitable for readers with moderate background in neuroscience)

Grafton, S. T., & Hamilton, A. F. d. C. (2007). Evidence for a distributed hierarchy of action representation in the brain. *Human Movement Science*, *26*(4), 590–616. doi:https://doi.org/10.1016/j.humov.2007.05.009

(Reviews behavioral and neural evidence to argue in favor of a functional-anatomic hierarchy in the brain supporting action planning and execution; suitable for readers with minimal background in neuroscience)

Gross, J. (2019). Magnetoencephalography in cognitive neuroscience: A primer. *Neuron*, *104*(2), 189–204.

(Describes magnetoencephalography and discusses its strengths and limitations; suitable for readers with minimal background in neuroscience)

Gur, M., & Snodderly, D. M. (1997). A dissociation between brain activity and perception: Chromatically opponent cortical neurons signal chromatic flicker that is not perceived. *Vision Research*, *37*(4), 377–382. doi:10.1016/s0042-6989(96)00183-6

(A neuroscience experiment that exemplifies how the brain can register visual information that is not perceived consciously; suitable for readers with moderate background in neuroscience)

Haggard, P. (2019). The neurocognitive bases of human volition. *Annual Review of Psychology*, *70*(1), 9–28. doi:10.1146/annurev-psych-010418-103348

(A comprehensive overview of volition and its neural basis that describes essential features of volition and reviews related research in psychology and neuroscience; suitable for readers with minimal background in philosophy, psychology, and neuroscience)

Haggard, P., Clark, S., & Kalogeras, J. (2002). Voluntary action and conscious awareness. *Nature Neuroscience*, *5*(4), 382–385. doi:10.1038/nn827

(Demonstrates that human subjects perceive voluntary actions as occurring later and sensory consequences as occurring earlier than actuality (coined "intentional binding") and argues that distinct neural mechanisms produce this effect in conscious awareness; suitable for readers with minimal background in neuroscience)

Haggard, P., & Eimer, M. (1999). On the relation between brain potentials and the awareness of voluntary movements. *Experimental Brain Research*, *126*(1), 128–133. doi:10.1007/s002210050722

(Using EEG while human subjects performed voluntary movements, suggests that the lateralized readiness potential may cause awareness of movement initiation; suitable for readers with moderate background in neuroscience)

Haken, H. (1991). *Synergetic Computers and Cognition: A Top-Down Approach to Neural Nets.* Springer-Verlag.

(Uses the mathematical and conceptual tools of synergetics to present a novel approach to neural nets, the "synergetic computer," and thus offers an alternative to "neuro-computers"; suitable for readers with minimal background in neuroscience)

Hallett, M. (2016). Physiology of free will. *Annals of Neurology*, *80*(1), 5–12.

(Describes free will as a perception that is not fully processed until after movement, reviewing neurophysiological findings related to the senses of willing and agency; suitable for readers with minimal background in neuroscience)

Hanakawa, T., Goldfine, A. M., & Hallett, M. (2017). A common function of basal ganglia-cortical circuits subserving speed in both motor and cognitive domains. *eNeuro, 4*(6).

(Uses fMRI in healthy human subjects and Parkinson's disease patients to highlight the role of basal ganglia-thalamo-cortical circuits in cognitive and motor agility; suitable for readers with moderate background in neuroscience)

Hare, T. A., Camerer, C. F., & Rangel, A. (2009). Self-control in decision-making involves modulation of the vmPFC valuation system. *Science*, *324*(5927), 646–648. doi:10.1126/science.1168450

(Uses an fMRI study of dieters engaging in food consumption decisions to propose that goal-directed decisions have their basis in a common value signal encoded in ventromedial prefrontal cortex and that exercising self-control involves the modulation of this value signal by dorsolateral prefrontal cortex; suitable for readers with minimal background in neuroscience)

Hassin, R. R. (2013). Yes it can: On the functional abilities of the human unconscious. *Perspectives on Psychological Science, 8*(2), 195–207. doi:10.1177/1745691612460684

(Argues that unconscious processes can perform the same fundamental, high-level functions that conscious processes can perform by considering evolutionary pressures and the availability of mental resources, as well as evidence from various subfields of the cognitive sciences; suitable for readers with moderate background in neuroscience)

Haynes, J.-D. (2011). Decoding and predicting intentions. *Annals of the New York Academy of Sciences, 1224*(1), 9–21. doi:10.1111/j.1749-6632.2011.05994.x

(Discusses what conclusions can be drawn from results suggesting that actions can be predicted from brain activity several seconds before subjects report deciding; suitable for readers with moderate background in neuroscience)

He, S., & MacLeod, D. I. A. (2001). Orientation-selective adaptation and tilt after-effect from invisible patterns. *Nature, 411*(6836), 473–476. doi:10.1038/35078072

(A neuroscience experiment that exemplifies how the brain can register visual information that is not perceived consciously; suitable for readers with moderate background in neuroscience)

Hieronymi, P. (2008). Responsibility for believing. *Synthese, 161*(3), 357–373.

(Proposes an account according to which beliefs, which seem to be involuntary, are a central example of the sort of thing for which we are most fundamentally responsible; suitable for readers with moderate background in philosophy)

Hieronymi, P. (2011). Reasons for action. *Proceedings of the Aristotelian Society, 111*, 407–427. doi:10.1111/j.1467-9264.2011.00316.x

(Responds to Donald Davidson's account of the relation between a reason and an action by offering an alternative account of the agent's own reasons for acting; suitable for readers with moderate background in philosophy)

Hieronymi, P. (2014). Reflection and responsibility. *Philosophy and Public Affairs, 42*(1), 3–41.

(Argues that the question of responsibility cannot be answered by appeal to self-awareness; suitable for readers with moderate background in philosophy)

Hieronymi, P. (forthcoming). Reasoning first. In R. Chang & K. Sylvan (Eds.), *Routledge Handbook of Practical Reasoning*. Routledge.

(Argues that all three kinds of reasons—epistemic, practical, and explanatory—share in common that they play a certain role in reasoning; suitable for readers with moderate background in philosophy)

Hieronymi, P. (in progress). Minds that matter.
  (Takes up traditional problems about free will and moral responsibility
  and argues that they can be avoided; suitable for readers with moderate
  background in philosophy)
Hiser, J., & Koenigs, M. (2018). The multifaceted role of the ventromedial prefrontal cortex in emotion, decision making, social cognition, and psychopathology. *Biological Psychiatry*, *83*(8), 638–647. doi:https://doi.org/10.1016/
  j.biopsych.2017.10.030
  (Summarizes data from a diverse collection of human and animal studies
  demonstrating that the ventromedial prefrontal cortex is a key node of cortical and subcortical networks that subserve at least three broad domains of
  psychological function linked to psychopathology; suitable for readers with
  moderate background in neuroscience)
Hitchcock, C. (2018). Probabilistic causation. In E. N. Zalta (Ed.), *The Stanford
  Encyclopedia of Philosophy* (Fall 2018 ed.). https://plato.stanford.edu/
  entries/causation-probabilistic/
  (A detailed discussion of the distinction between determinism and causation and their compatibility with freedom; suitable for readers with moderate background in philosophy)
Hobbes, T. (1651/1982). *Leviathan*. Penguin Classics.
  (A famous work concerning the structure of society and legitimate government, including a compatibilist account of choice and free will; suitable
  for readers with minimal background in philosophy)
Hoffstaedter, F., Grefkes, C., Zilles, K., & Eickhoff, S. B. (2013). The "what" and
  "when" of self-initiated movements. *Cerebral Cortex*, *23*(3), 520–530.
  (Uses fMRI in human subjects making finger movements, with the
  "what" (right or left) vs. "when" components either self-chosen or externally
  cued to show different condition combinations recruit distinct networks;
  suitable for readers with moderate background in neuroscience)
Horgan, T., Tienson, J., & Graham, G. (2003). The phenomenology of first-
  person agency. In S. Walter & H. D. Heckmann (Eds.), *Physicalism and Mental
  Causation: The Metaphysics of Mind and Action* (pp. 323–340). Academic.
  (Discusses philosophical issues that arise from the experience of acting
  voluntarily, focusing specifically on the aspect of self-as-source, the aspect
  of purposiveness, and the aspect of voluntariness; suitable for readers with
  minimal background in philosophy)
Huxley, A. (1932). *Brave New World*. Chatto & Windus.
  (A famous work of science fiction about a technologically advanced future where humans are genetically bred, socially indoctrinated, and phar-
  maceutically anesthetized to uphold an authoritarian ruling order; suitable
  for all readers)
Jeannerod, M. (1997). *The Cognitive Neuroscience of Action*. Blackwell.

(Provides a concise, accessible review of cognitive neuroscience research on the nature and role of different representations in the planning and execution of movements; suitable for readers with minimal background in neuroscience)

Kandel, E. R., Schwartz, J. H., Jessell, T. M., Siegelbaum, S. A., & Hudspeth, A. J. (2012). *Principles of Neural Science* (5th ed.). McGraw-Hill Education.

(Prominent researchers survey fields of research, spanning the entire spectrum of neural science; suitable for readers with minimal background in neuroscience)

Kane, R. (1996). *The Significance of Free Will.* Oxford University Press.

(Disagrees with the common view that free will is a system-level or "top-down" form of control; suitable for readers with moderate background in philosophy)

Kane, R. (2007). Libertarianism. In J. M. Fischer, R. Kane, D. Pereboom, & M. Vargas (Eds.), *Four Views on Free Will* (pp. 5–44). Blackwell.

(An overview of the libertarian view of free will, which claims that people have free will even though causal determinism is incompatible with freedom of action; suitable for readers with minimal background in philosophy)

Khalighinejad, N., Brann, E., Dorgham, A., & Haggard, P. (2019). Dissociating cognitive and motoric precursors of human self-initiated action. *Journal of Cognitive Neuroscience, 31*(5), 754–767.

(Prompts subjects to view moving that sometimes move coherently to the right or left and then either report the movement direction of the dots or skip report and demonstrates that EEG across-trial variability decreases more markedly before self-initiated compared with externally triggered skip actions, especially if the skip responses recruited deliberate planning; suitable for readers with moderate background in neuroscience)

Kim, C.-Y., & Blake, R. (2005). Psychophysical magic: Rendering the visible "invisible." *Trends in Cognitive Sciences, 9*(8), 381–388. doi:https://doi.org/10.1016/j.tics.2005.06.012

(Reviews experimental strategies for dissociating the presentation of stimuli from conscious awareness; suitable for readers with minimal background in neuroscience)

King, J.-R., Sitt, Jacobo D., Faugeras, F., Rohaut, B., El Karoui, I., Cohen, L., . . . Dehaene, S. (2013). Information sharing in the brain indexes consciousness in noncommunicative patients. *Current Biology, 23*(19), 1914–1919. doi:https://doi.org/10.1016/j.cub.2013.07.075

(Reports that a novel measure of information sharing increases with consciousness state and that this effect distinguishes patients in vegetative state, minimally conscious state, and conscious state, thereby supporting distributed theories of conscious processing and opening up the possibility of an automatic detection of conscious states; suitable for readers with strong background in neuroscience)

Koch, C. (2004). *The Quest for Consciousness: A Neurobiological Approach.* Roberts & Co.

(An exploration of the neurological explanation for consciousness based on the author's research on vision and perception; suitable for readers with minimal background in neuroscience)

Koch, C., Massimini, M., Boly, M., & Tononi, G. (2016). Neural correlates of consciousness: progress and problems. *Nature Reviews Neuroscience, 17*(5), 307–321. doi:10.1038/nrn.2016.22

(Summarizes neuroscientific evidence suggesting that representations at the level of the lateral geniculate nucleus and V1 are not part of the neural basis of consciousness and that the best candidates for full and content-specific correlates of consciousness are located in a temporo-parietal-occipital hot zone; suitable for readers with moderate background in neuroscience)

Koechlin, E., Ody, C., & Kouneiher, F. (2003). The architecture of cognitive control in the human prefrontal cortex. *Science, 302*(5648), 1181–1185. doi:10.1126/science.1088545

(Using fMRI in humans, reports that the lateral prefrontal cortex is organized as a cascade of executive processes important for controlling action in accordance with perceptual content; suitable for readers with moderate background in neuroscience)

Kornhuber, H., & Deecke, L. (1965). Changes in the brain potential in voluntary movements and passive movements in man: Readiness potential and reafferent potentials. *Pflugers Archiv fur Die Gesamte Physiologie des Menschen und der Tiere, 284*, 1–17.

(Uses EEG while human subjects performed voluntary or passive movements to identify accompanying slow brain potentials and establish the readiness potential; suitable for readers with minimal background in neuroscience)

Kranick, S. M., & Hallett, M. (2013). Neurology of volition. *Experimental Brain Research, 229*(3), 313–327. doi:10.1007/s00221-013-3399-2

(Reviews neurological disorders related to volition and discusses how these disorders may illuminate the neural basis of percepts of volition; suitable for readers with minimal background in neuroscience)

Kreiman, G. (2021). *Biological and Computer Vision.* Cambridge University Press.

(Surveys the neuroscientific study of neuronal computations in visual cortex as well as the field of biologically-inspired artificial intelligence; suitable for readers with moderate background in neuroscience)

Kuhn, T. S. (1962). *The Structure of Scientific Revolutions.* University of Chicago Press.

(Surveys the history of science, analyzes scientific revolutions, and suggests that science occurs in cycles that repeat over time; suitable for readers with minimal background in philosophy and minimal background in neuroscience)

Kuzmanovic, B., Rigoux, L., & Tittgemeyer, M. (2018). Influence of vmPFC on dmPFC predicts valence-guided belief formation. *Journal of Neuroscience*, *38*(37), 7996–8010. doi:10.1523/jneurosci.0266-18.2018

(Using fMRI in human subjects and computational causal modeling, demonstrates that activation of the ventromedial prefrontal cortex is associated with value-based belief formation and updating; suitable for readers with moderate background in neuroscience)

Lamme, V. A. F. (2003). Why visual attention and awareness are different. *Trends in Cognitive Sciences, 7*(1), 12–18. doi:https://doi.org/10.1016/S1364-6613(02)00013-X

(An argument for the recurrent processing approach to consciousness, which dictates that conscious processing occurs well after stimuli are extensively processed in the visual centers of the brain; suitable for readers with moderate background in neuroscience)

Lau, H. C., Rogers, R. D., & Passingham, R. E. (2007). Manipulating the experienced onset of intention after action execution. *Journal of Cognitive Neuroscience, 19*(1), 81–90. doi:10.1162/jocn.2007.19.1.81%M17214565

(A TMS experiment that suggests the consciousness of a decision can occur at least in part after the action; suitable for readers with minimal background in philosophy and moderate background in neuroscience)

Libet, B. (1985). Unconscious cerebral initiative and the role of conscious will in voluntary action. *Behavioral and Brain Sciences, 8*(4), 529–539. doi:10.1017/S0140525X00044903

(A seminal neuroscience study that concludes that conscious will does not initiate voluntary acts, as electrophysiological readiness potentials precede reports of having decided to move; suitable for readers with minimal background in neuroscience)

Libet, B., Gleason, C. A., Wright, E. W., & Pearl, D. K. (1983). Time of conscious intention to act in relation to onset of cerebral activity (readiness-potential): The unconscious initiation of a freely voluntary act. *Brain, 106* (Pt 3), 623–642. doi:10.1093/brain/106.3.623

(A seminal neuroscience paper that concludes that the onset of the readiness potential in relation to the onset of the reported urge to move demonstrates that we do not have free will; suitable for readers with minimal background in neuroscience)

Liljenström, H. (2016). Multi-scale causation in brain dynamics. In R. Kozma & W. J. Freeman (Eds.), *Cognitive Phase Transitions in the Cerebral Cortex— Enhancing the Neuron Doctrine by Modeling Neural Fields* (pp. 177–186). Springer.

(Presents both upward and downward causation in cortical neural systems using computational methods with focus on cortical fluctuations and discusses philosophical implications from these studies; suitable for readers with strong background in neuroscience and moderate background in philosophy)

Liljenström, H. (2018). Intentionality as a driving force. *Journal of Consciousness Studies*, 25(1–2), 206–229.

(Reflects upon Walter Freeman's work on intentionality and its relation to mesoscopic dynamics, elaborates on the role of intentionality for decision-making and free will, describes a computational model of decision-making, and discusses the intention and attention aspects of consciousness; suitable for readers with moderate background in philosophy and neuroscience)

Liljenström, H., & Hasselmo, M. (1995). Cholinergic modulation of cortical oscillatory dynamics. *Journal of Neurophysiology*, 74(1), 288–297. doi:10.1152/jn.1995.74.1.288

(Using a computational model study, finds that cholinergic modulation may be involved in switching the dynamics of the olfactory cortex between those appropriate for learning and those appropriate for recall; suitable for readers with strong background in neuroscience)

Liu, C., Sun, Z., Jou, J., Cui, Q., Zhao, G., Qiu, J., & Tu, S. (2016). Unconscious processing of facial emotional valence relation: Behavioral evidence of integration between subliminally perceived stimuli. *PLoS One*, 11(9), e0162689. doi:10.1371/journal.pone.0162689

(Presents the first behavioral study exploring the integration between unconscious emotional stimuli using a modified priming paradigm and concludes that integration between different unconsciously perceived stimuli can occur; suitable for readers with minimal background in neuroscience)

Locke, J. (1689/1975). *An Essay concerning Human Understanding* (P. H. Nidditch, Ed.). Oxford University Press.

(A wide-ranging work that includes the suggestion that freedom of action, which requires the ability to act otherwise, is not necessary for freedom of will or moral responsibility; suitable for readers with minimal background in philosophy)

Louail, M., Gilissen, E., Prat, S., Garcia, C., & Bouret, S. (2019). Refining the ecological brain: Strong relation between the ventromedial prefrontal cortex and feeding ecology in five primate species. *Cortex, 118*, 262–274. doi:https://doi.org/10.1016/j.cortex.2019.03.019

(Using a comparative approach with predictions coming both from behavioral ecology and cognitive neuroscience, provides evidence that feeding ecology played a key role in the development of specific cognitive skills, which rely upon the expansion of a specific cortical area; suitable for readers with moderate background in neuroscience)

Mai, J. K., Majtanik, M., & Paxinos, G. (2015). *Atlas of the Human Brain*. Academic Press.

(Presents the anatomy of the brain at macroscopic and microscopic levels; suitable for readers with minimal background in neuroscience)

Maimon, G., & Assad, J. A. (2006). A cognitive signal for the proactive timing of action in macaque LIP. *Nature Neuroscience, 9*(7), 948–955.

(Records single-neuron activity in macaques to identify neuronal activity in the lateral intraparietal area corresponding to internally generated arm movements; suitable for readers with moderate background in neuroscience)

Maniscalco, B., & Lau, H. (2012). A signal detection theoretic approach for estimating metacognitive sensitivity from confidence ratings. *Consciousness and Cognition, 21*(1), 422–430. doi:https://doi.org/10.1016/j.concog.2011.09.021

(Introduces a novel measure called meta-d′ to show that subjects' metacognitive sensitivity— i.e., the efficacy with which their confidence ratings discriminate between their correct and incorrect stimulus classifications— is close to, but significantly below, optimality; suitable for readers with minimal background in neuroscience)

Maoz, U., Yaffe, G., Koch, C., & Mudrik, L. (2019). Neural precursors of decisions that matter—An ERP study of deliberate and arbitrary choice. *ELife, 8*, e39787.

(Using EEG while human subjects made arbitrary or deliberate decisions about donations to nonprofit organizations, demonstrates that the readiness potential is present for arbitrary but not for deliberate decisions; suitable for readers with moderate background in neuroscience)

Matsuhashi, M., & Hallett, M. (2008). The timing of the conscious intention to move. *European Journal of Neuroscience, 28*(11), 2344–2351.

(Using EEG and a novel probe paradigm while subjects performed or vetoed self-paced movements, demonstrates that the onset of intention occurs much earlier than in the Libet experiments, but still after the onset of readiness potential; suitable for readers with moderate background in neuroscience)

Mele, A. R. (1995). *Autonomous Agents: From Self Control to Autonomy.* Oxford University Press.

(Argues for a historical dimension to free will, which requires that the process whereby we become who we are is psychologically "integrated," constituting a reasonably coherent unfolding narrative; suitable for readers with moderate background in philosophy)

Mele, A. R. (2014). *Free: Why Science Hasn't Disproved Free Will.* Oxford University Press.

(Explores the implications of recent research in neuroscience and psychology for philosophical debates about free will; suitable for readers with minimal background in philosophy)

Monod, J. (1971). *Chance and Necessity: An Essay on the Natural Philosophy of Modern Biology* (Vol. 1). Knopf.

(Relies on genetics and molecular biology to reject the "animist" conception of man that has dominated Western thought and argues that life is only the result of natural processes by "pure chance" and that scientific knowledge denies the concepts of destiny and evolutionary purpose; suitable for readers with moderate background in philosophy)

Morsella, E., Godwin, C. A., Jantz, T. K., Krieger, S. C., & Gazzaley, A. (2016). Homing in on consciousness in the nervous system: An action-based synthesis. *Behavioral and Brain Sciences, 39,* E168.

(Presents passive frame theory, which suggests that the function of consciousness in the somatic nervous system is to constrain and direct skeletal muscle output; suitable for readers with moderate background in neuroscience)

Mudrik, L., Breska, A., Lamy, D., & Deouell, L. Y. (2011). Integration without awareness: Expanding the limits of unconscious processing. *Psychological Science, 22*(6), 764–770. doi:10.1177/0956797611408736

(A binocular rivalry experiment that suggests visual awareness is not needed for object-background integration or for processing the likelihood of an object to appear within a given semantic context, but may be needed for dealing with novel situations; suitable for readers with moderate background in psychology)

Mudrik, L., Faivre, N., & Koch, C. (2014). Information integration without awareness. *Trends in Cognitive Sciences, 18*(9), 488–496. doi:10.1016/j.tics.2014.04.009

(Examines experimental literature on perceptual and cognitive integration of spatiotemporal, multisensory, semantic, and novel information and finds that, while some integrative processes can occur without awareness, their scope is limited to smaller integration windows, to simpler associations, or to ones that were previously acquired consciously; suitable for readers with moderate background in psychology and neuroscience)

Mudrik, L., & Maoz, U. (2014). "Me & my brain": Exposing neuroscience's closet dualism. *Journal of Cognitive Neuroscience, 27*(2), 211–221.

(Discusses the origins and implications of the "double-subject fallacy" in neuroscience, which treats the brain and the entire person as two independent subjects who can occupy divergent psychological states and have complex interactions with each other; suitable for readers with minimal background in philosophy and neuroscience)

Muhammed, K., Manohar, S., & Husain, M. (2015). Mechanisms underlying apathy in Parkinson's disease. *The Lancet, 385,* S71. doi:10.1016/S0140-6736(15)60386-5

(Using eye-tracking and pupillometry in healthy subjects and subjects with Parkinson's disease, demonstrates that saccades and pupil dilation are affected by reward and may thus be used as measures of motivation and apathy; suitable for readers with minimal background in neuroscience)

Murray, S., Murray, E. D., Stewart, G., Sinnott-Armstrong, W., & De Brigard, F. (2019). Responsibility for forgetting. *Philosophical Studies, 176*(5), 1177–1201.

(Provides evidence that people tend to hold others responsible for some forgetfulness and lack of vigilance, suggesting that there can be responsibility without occurrent beliefs or intentions; suitable for readers with minimal background in philosophy)

Nahmias, E., Morris, S., Nadelhoffer, T., & Turner, J. (2005). Surveying freedom: Folk intuitions about free will and moral responsibility. *Philosophical Psychology, 18*(5), 561–584. doi:10.1080/09515080500264180
(A seminal experimental philosophy paper that surveys non-philosophers' views on free will in a determined world and other scenarios; suitable for readers with minimal background in philosophy)

Nazir, A. H., & Liljenström, H. (2015). A cortical network model of cognitive and emotional influences in human decision making. *Biosystems, 136,* 128–141. doi:https://doi.org/10.1016/j.biosystems.2015.07.004
(Presents a model of decision-making involving three neural structures (amygdala, OFC, and LPFC) to illuminate the emotional and cognitive processes involved; suitable for readers with moderate background in neuroscience)

Newell, B. R., & Shanks, D. R. (2014). Unconscious influences on decision making: A critical review. *Behavioral and Brain Sciences, 37*(1), 1–19.
(Employs a novel framework to evaluate the evidence in favor of the popular claim that unconscious influences play a large causal role on behavior and concludes that these influences have been ascribed inflated and erroneous explanatory power in theories of decision-making; suitable for readers with minimal background in philosophy and neuroscience)

Nichols, S. (2014). The indeterminist intuition: Source and status. *The Monist, 95*(2), 290–307. doi:10.5840/monist201295216
(Argues that people's common belief that their choices are not determined is unjustified, because it depends on a presumption that we know the factors that influence our decision-making, which is undermined by recent work in cognitive science; suitable for readers with minimal background in philosophy)

Noble, D. (2008). *The Music of Life: Biology beyond Genes.* Oxford University Press.
(Introduces readers to the increasingly popular discipline of systems biology and argues that, contrary to Dawkins's suggestion, understanding life requires looking beyond the "selfish gene" to consider a much wider variety of interacting biological levels; suitable for all readers)

O'Connor, T. (2019). How do we know that we are free? *European Journal of Analytic Philosophy, 15*(2), 79–98.
(Explores the question of whether the existence of our limited freedom can be known prior to scientific investigation; suitable for readers with moderate background in philosophy)

O'Connor, T. (2020). Emergent properties. In E. N. Zalta (Ed.), *The Stanford Encyclopedia of Philosophy* (Fall 2020 ed.). https://plato.stanford.edu/entries/properties-emergent/
(A detailed analysis of emergent properties and emergence, discussing weak vs. strong emergence; suitable for readers with moderate background in philosophy)

O'Connor, T., & Franklin, C. (2020). Free will. In E. N. Zalta (Ed.), *The Stanford Encyclopedia of Philosophy* (Fall 2020 ed.). https://plato.stanford.edu/entries/freewill/
(A comprehensive discussion of the issues raised by the concept of free will and the ways that philosophers have theorized about them; suitable for readers with minimal background in philosophy)

Ott, T., & Nieder, A. (2019). Dopamine and cognitive control in prefrontal cortex. *Trends in Cognitive Sciences, 23*(3), 213–234. doi:https://doi.org/10.1016/j.tics.2018.12.006
(Reviews recent research and concludes that dopamine receptors in prefrontal cortex control three key aspects of cognitive control: gating, maintaining, and relaying; suitable for readers with moderate background in neuroscience)

Pacherie, E. (2008). The phenomenology of action: A conceptual framework. *Cognition, 107*(1), 179–217. doi:https://doi.org/10.1016/j.cognition.2007.09.003
(Discusses the structures of experience and consciousness related to action and presents a three-tiered dynamic model of intention and a theory of action control; suitable for readers with minimal background in philosophy and moderate background in psychology)

Palmer, T. D., & Ramsey, A. K. (2012). The function of consciousness in multisensory integration. *Cognition, 125*(3), 353–364. doi:https://doi.org/10.1016/j.cognition.2012.08.003
(Explores the function of consciousness in multisensory integration, specifically during audio-visual speech, cross-modal visual attention guidance, and McGurk cross-modal integration, and finds that cross-modal effects can occur unconsciously but that the influencing modality must be consciously perceived; suitable for readers with moderate background in psychology)

Parés-Pujolràs, E., Travers, E., Ahmetoglu, Y., & Haggard, P. (2021). Evidence accumulation under uncertainty—a neural marker of emerging choice and urgency. *NeuroImage, 232.*
(Participants decide whether or not to act while monitoring discrete visual stimuli that contain either strong or ambiguous evidence to conclude that the P3 EEG component tracks the evolution of a decision and that the readiness potential proceeds internally and exogenously initiated actions; suitable for readers with moderate background in neuroscience)

Passingham, R. E. (1987). Two cortical systems for directing movement. *Ciba Foundation Symposium, 132,* 151–164. doi:10.1002/9780470513545.ch10

(Explores action-relevant neural correlates in macaques to suggest that premotor areas direct action based on visual cues and the supplementary motor area directs actions based on proprioceptive cues; suitable for readers with minimal background in neuroscience)

Passingham, R. E., & Lau, H. (2006). Free choice and the human brain. In S. Pockett, W. Banks, & S. Gallagher (Eds.), *Does Consciousness Cause Behavior?* (pp. 53–72). MIT Press.

(Surveys neuroscientific studies to argue that phenomenally conscious decisions are "epiphenomenal" or have no causal effects on bodily movements; suitable for readers with minimal background in neuroscience)

Peirce, C. S., & Jastrow, J. (1884). On small differences in sensation. *Memoirs of the National Academy of Sciences, 3,* 73–83.

(Presents the seminal claim for unconscious perception using evidence from psychology; suitable for readers with moderate background in psychology)

Peters, M. A. K., Kentridge, R. W., Phillips, I., & Block, N. (2017). Does unconscious perception really exist? Continuing the ASSC20 debate. *Neuroscience of Consciousness, 2017*(1). doi:10.1093/nc/nix015

(Includes four short essays on unconscious perception by empirical scientists studying consciousness; Block's essay explores the methodological possibility of "shaving off" the conscious part of a perception to study the unconscious perception within; suitable for readers with moderate background in neuroscience)

Pitts, M. A., Metzler, S., & Hillyard, S. A. (2014). Isolating neural correlates of conscious perception from neural correlates of reporting one's perception. *Frontiers in Psychology, 5*(1078). doi:10.3389/fpsyg.2014.01078

(Provides evidence for the recurrent processing approach to consciousness, which dictates that conscious processing occurs well after stimuli are extensively processed in the visual centers of the brain; suitable for readers with moderate background in neuroscience)

Popper, K. (1934/2005). *The Logic of Scientific Discovery*: Routledge.

(Seminal work in which philosopher of science Karl Popper argues that science should adopt a methodology based on falsifiability; suitable for readers with minimal background in philosophy)

Reingold, E. M., & Merikle, P. M. (1988). Using direct and indirect measures to study perception without awareness. *Perception & Psychophysics, 44*(6), 563–575. doi:10.3758/BF03207490

(Suggests comparing the sensitivity of direct and indirect measures of perception to evaluate perception without awareness; suitable for readers with moderate background in neuroscience)

Robb, D., & Heil, J. (2019). Mental causation. In E. N. Zalta (Ed.), *The Stanford Encyclopedia of Philosophy* (Summer 2019 ed.). https://plato.stanford.edu/entries/mental-causation/
(A detailed analysis of whether mental states, events, or properties can cause physical effects; suitable for readers with moderate background in philosophy)

Romo, R., & Schultz, W. (1992). Role of primate basal ganglia and frontal cortex in the internal generation of movements. III: Neuronal activity in the supplementary motor area. *Experimental Brain Research, 91*(3), 396.
(Recording single neurons in primates completing externally instructed or self-initiated movements, suggests a conjoint role for the supplementary motor area and the striatum; suitable for readers with moderate background in neuroscience)

Rosenthal, D. M. (1986). Two concepts of consciousness. *Philosophical Studies, 49*(3), 329–359. doi:10.1007/BF00355521
(Distinguishes between two stances on the consciousness of mental states and argues that the higher-order-thought theory of consciousness, which allows for substantial unconscious processing prior to conscious processing, is better suited to accommodate empirical evidence; suitable for readers with moderate background in philosophy)

Roskies, A. (2006). Neuroscientific challenges to free will and responsibility. *Trends in Cognitive Sciences, 10*(9), 419–423. doi:10.1016/j.tics.2006.07.011
(Argues that neuroscience by itself can never settle whether anyone has free will since it cannot reveal whether brains are deterministic or incorporate random processes; suitable for readers with minimal background in philosophy)

Roskies, A. (2011). Why Libet's studies don't pose a threat to free will. In W. Sinnott-Armstrong (Ed.), *Conscious Will and Responsibility: A Tribute to Benjamin Libet* (pp. 11–22). Oxford University Press.
(Argues that Libet's original argument is flawed and does not prove that we do not have free will; suitable for readers with minimal background in philosophy and neuroscience)

Roskies, A. (2014). Can neuroscience resolve issues about free will? In W. Sinnott-Armstrong (Ed.), *Moral Psychology. Volume 4: Free Will & Moral Responsibility*. MIT Press.
(Discusses single-neuron recordings and classic experiments on perceptual decision-making to explore whether brain processes are stochastic, and sides with compatibilism over libertarianism; suitable for readers with minimal background in philosophy and neuroscience)

Roskies, A., & Nichols, S. (2008). Bringing moral responsibility down to earth. *Journal of Philosophy, 105*(7), 371–388. doi:10.2307/20620111
(Surveys and explains non-philosophers' divergent views on moral responsibility in various hypothetical worlds; suitable for readers with minimal background in philosophy)

Schlegel, A., Alexander, P., Sinnott-Armstrong, W., Roskies, A., Tse, P. U., & Wheatley, T. (2013). Barking up the wrong free: Readiness potentials reflect processes independent of conscious will. *Experimental Brain Research*, *229*(3), 329–335. doi:10.1007/s00221-013-3479-3

(Using EEG while human subjects move voluntarily, argues that the readiness potential and the lateralized readiness potential reflect processes independent of conscious will; suitable for readers with minimal background in neuroscience)

Schneider, L., Houdayer, E., Bai, O., & Hallett, M. (2013). What we think before a voluntary movement. *Journal of Cognitive Neuroscience*, *25*(6), 822–829.

(Using an optimized EEG signal derived from multiple variables, predicts impending movements in real time on a single-trial basis and demonstrates that a subject's thoughts can be unrelated to movement while the brain is preparing for it; suitable for readers with moderate background in neuroscience)

Schrödinger, E. (1944). *What Is Life? The Physical Aspect of the Living Cell*. Cambridge University Press.

(An accessible and groundbreaking book that proved to be one of the spurs of the birth of molecular biology and the subsequent discovery of DNA; suitable for all readers)

Schultze-Kraft, M., Birman, D., Rusconi, M., Allefeld, C., Görgen, K., Dähne, S., . . . Haynes, J.-D. (2016). The point of no return in vetoing self-initiated movements. *Proceedings of the National Academy of Sciences*, *113*(4), 1080–1085. doi:10.1073/pnas.1513569112

(Trains a brain-computer interface to detect readiness potentials in real time while human subjects veto (stop) movements, finds vetoing is possible after RP onset but only up to 200 ms before movement; suitable for readers with moderate background in neuroscience)

Schurger, A. (2018). Specific relationship between the shape of the readiness potential, subjective decision time, and waiting time predicted by an accumulator model with temporally autocorrelated input noise. *eneuro*, *5*(1), ENEURO.0302-0317.2018. doi:10.1523/ENEURO.0302-17.2018

(Tests various extensions of the leaky stochastic-accumulator model that provides an alternative explanation for the readiness potential in voluntary action; suitable for readers with strong background in neuroscience)

Schurger, A., Sarigiannidis, I., Naccache, L., Sitt, J. D., & Dehaene, S. (2015). Cortical activity is more stable when sensory stimuli are consciously perceived. *Proceedings of the National Academy of Sciences*, *112*(16), E2083–E2092. doi:10.1073/pnas.1418730112

(Presents a new experimental design that records neural stability at the single-trial level, which can be used to discriminate the conscious state of brain-injured patients and which validates the relevance of transient neural stability for conscious perception; suitable for readers with strong background in neuroscience)

Schurger, A., Sitt, J. D., & Dehaene, S. (2012). An accumulator model for spontaneous neural activity prior to self-initiated movement. *Proceedings of the National Academy of Sciences, 109*(42), E2904–E2913. doi:10.1073/pnas.1210467109

(Suggests that the readiness potential might reflect activity in a leaky stochastic-accumulator model rather than reflecting the early onset of a decision to move; suitable for readers with strong background in neuroscience)

Schurger, A., & Uithol, S. (2015). Nowhere and everywhere: The causal origin of voluntary action. *Review of Philosophy and Psychology, 6*(4), 761–778. doi:10.1007/s13164-014-0223-2

(Suggests that intentions might emerge from a distributed network in the brain rather than from a locus of intention; suitable for readers with moderate background in neuroscience)

Schurger A., Hu P., Pak J., & Roskies A. L. (2021). What is the readiness potential? *Trends in Cognitive Sciences, 25*(7), 558–570.

(Argues recent computational work on the readiness potential (RP) calls for a reassessment of its relevance for understanding volition and free will; suitable for readers with moderate background in neuroscience)

Scott, S. H. (2004). Optimal feedback control and the neural basis of volitional motor control. *Nature Reviews Neuroscience, 5*(7), 532–545.

(Argues optimal feedback control theory accounts for adaptive and controlled movement by linking levels of the motor system (motor behavior, limb mechanics, and neural control); suitable for readers with moderate background in neuroscience)

Searle, J. R. (1980). Minds, brains, and programs. *Behavioral and Brain Sciences, 3*(3), 417–424. doi:10.1017/S0140525X00005756

(Proposes the famous Chinese Room thought experiment as evidence that the kind of proficiency in converting inputs to outputs shown by an entity passing the Turing test does not imply intentionality; suitable for readers with moderate background in philosophy and minimal background in neuroscience)

Seth, A. K., Dienes, Z., Cleeremans, A., Overgaard, M., & Pessoa, L. (2008). Measuring consciousness: Relating behavioural and neurophysiological approaches. *Trends in Cognitive Sciences, 12*(8), 314–321. doi:https://doi.org/10.1016/j.tics.2008.04.008

(Reviews behavioral and brain-based methods for measuring consciousness and how they conflict, along with touching on existing theories about consciousness; suitable for readers with minimal background in psychology and neuroscience)

Seth, A. K., Verschure, P. F., Morsella, E., O'Regan, J. K., Blanke, O., Butz, M. V., . . . Kyselo, M. (2016). Action-oriented understanding of consciousness and the structure of experience. In *The Pragmatic Turn: Toward Action-Oriented Views in Cognitive Science* (pp. 261–282). MIT Press.

(Combines interdisciplinary perspectives on an action-oriented approach to cognitive science, discussing how actions shape consciousness, consciousness of action, self-experience, and relevant theoretical frameworks; suitable for readers with moderate background in neuroscience)

Shepherd, J. (2015). Conscious control over action. *Mind & Language, 30*(3), 320–344.

(Describes an understanding of conscious control that is compatible with models of overt action control and that allows consciousness to play a role in action; suitable for readers with minimal background in neuroscience)

Sherrington, C. (1952). *The Integrative Action of the Nervous System.* Cambridge University Press Archive.

(Book of lectures on action production and related processes in the nervous system; suitable for readers with moderate background in neuroscience)

Shibasaki, H., & Hallett, M. (2006). What is the Bereitschaftspotential? *Clinical Neurophysiology, 117*(11), 2341–2356.

(Reviews the Bereitschaftspotential, or readiness potential, its physiological source, components, influencing factors, role in voluntary movement, and variation in disorders of volition; suitable for readers with minimal background in neuroscience)

Silvanto, J., Cowey, A., Lavie, N., & Walsh, V. (2005). Striate cortex (V1) activity gates awareness of motion. *Nature Neuroscience, 8*(2), 143–144.

(A neuroscience experiment that suggests that feedback to V1 from higher visual areas is essential to visual awareness; suitable for readers with minimal background in neuroscience)

Silver, N. (2012). *The Signal and the Noise: Why So Many Predictions Fail—but Some Don't.* Penguin.

(Drawing on the author's own work forecasting baseball performance and political outcomes, examines the world of prediction and investigates how to distinguish a true signal from a universe of data; suitable for all readers)

Sinnott-Armstrong, W. (2012a). A case study in neuroscience and responsibility. In J. E. Fleming & S. Levinson (Eds.), *Evolution and Morality: NOMOS LII* (pp. 194–211). New York University Press.

(Uses a case study of a patient with a brain tumor to explore whether and when neuroscience can illuminate whether an agent is responsible; suitable for readers with minimal background in philosophy)

Sinnott-Armstrong, W. (2012b). Free contrastivism. In M. Blaauw (Ed.), *Contrastivism in Philosophy: New Perspectives* (pp. 134–153). Routledge.

(Proposes a compromise position between incompatibilist and compatibilist views of freedom and outlines which problems each view is suited to address; suitable for readers with minimal background in philosophy)

Sinnott-Armstrong, W. (2016). My brain made me do it—So what? In D. Edmonds (Ed.), *Philosophers Take On the World* (pp. 147–149). Oxford University Press.

(Uses a hypothetical dialogue between friends to explore why the fact that our brains cause our actions is not always sufficient to excuse those actions; suitable for all readers)

Sinnott-Armstrong, W. (2018). *Think Again: How to Reason and Argue.* Penguin.

(Identifies the components of good arguments as well as fallacies to avoid and demonstrates what good arguments can accomplish; suitable for all readers)

Sitt, J. D., King, J.-R., El Karoui, I., Rohaut, B., Faugeras, F., Gramfort, A., . . . Naccache, L. (2014). Large scale screening of neural signatures of consciousness in patients in a vegetative or minimally conscious state. *Brain, 137*(8), 2258–2270. doi:10.1093/brain/awu141

(Based on a review of previous experiments and current theories, identifies measures of the efficiency of EEG markers to differentiate patients in a vegetative state from those in a minimally conscious or conscious state; suitable for readers with moderate background in neuroscience)

Splittgerber, R. (2019). *Snell's Clinical Neuroanatomy* (8th ed.). Wolters Kluwer.

(Provides a clinically orientated understanding of neuroanatomy; suitable for all readers)

Sripada, C. (2016). Self-expression: A deep self theory of moral responsibility. *Philosophical Studies, 173*(5), 1203–1232. doi:10.1007/s11098-015-0527-9

(A compatibilist argument that free actions must be related to the agent's deep self; suitable for readers with moderate background in philosophy)

Steinbeis, N., Haushofer, J., Fehr, E., & Singer, T. (2014). Development of behavioral control and associated vmPFC–DLPFC connectivity explains children's increased resistance to temptation in intertemporal choice. *Cerebral Cortex, 26*(1), 32–42. doi:10.1093/cercor/bhu167

(Using behavioral and fMRI measures of choice tasks among 6–13 year-olds, shows developmental improvements in behavioral control to uniquely account for age-related changes in temporal discounting and that overcoming temptation during childhood occurs as a function of age-related increase in functional coupling between the vmPFC and the dlPFC; suitable for readers with moderate background in neuroscience)

Suzuki, K., Lush, P., Seth, A. K., & Roseboom, W. (2019). Intentional binding without intentional action. *Psychological Science, 30*(6), 842–853. doi:10.1177/0956797619842191

(Using virtual reality, demonstrates that subjects perceive causes as occurring later and subsequent effects as occurring earlier than actuality, whether or not they act intentionally; suitable for readers with moderate background in neuroscience)

Tang, H., Schrimpf, M., Lotter, W., Moerman, C., Paredes, A., Ortega Caro, J., . . . Kreiman, G. (2018). Recurrent computations for visual pattern completion. *Proceedings of the National Academy of Sciences, 115*(35), 8835–8840. doi:10.1073/pnas.1719397115

(Combines psychophysics, physiology, and computational models to test and present three pieces of evidence consistent with the hypothesis that pattern completion is implemented by recurrent computations; suitable for readers with moderate background in neuroscience)

Thompson, M. (2012). *Life and Action.* Harvard University Press.

(Argues that whether a person has an intention is a matter of whether certain sorts of questions are appropriately asked of her, so intentions are a social matter; suitable for readers with moderate background in philosophy)

Travers, E., Friedemann, M., & Haggard, P. (2021). The readiness potential reflects planning-based expectation, not uncertainty, in the timing of action. *Cognitive Neuroscience, 12*(1), 14–27.

(Uses a reinforcement learning paradigm where participants learned, through trial and error, the optimal time to act to demonstrate that the readiness potential amplitude grows with learning and thus reflects planning and anticipation rather than freedom from external constraint; suitable for readers with moderate background in neuroscience)

Travers, E., & Haggard, P. (2019). The readiness potential reflects internal source of actions, not decision uncertainty. *bioRxiv,* 782813. doi:10.1101/782813

(Uses EEG while subjects decide to act or not act in a gambling task, reports that the readiness potential is reflective of internally generated action rather than uncertainty; suitable for readers with moderate background in neuroscience)

Turing, A. (1950). Computing machinery and intelligence. *Mind, 59*(236), 433–460.

(A seminal thought experiment about whether machines can think that involves an "imitation game" where an interrogator asks questions to a machine and a human and tries to discern which is which based on their answers; suitable for readers with minimal background in philosophy)

Uithol, S., Burnston, D. C., & Haselager, P. (2014). Why we may not find intentions in the brain. *Neuropsychologia, 56,* 129–139. doi:https://doi.org/10.1016/j.neuropsychologia.2014.01.010

(Argues that the processes underlying action initiation and control are too dynamic and context-sensitive to support the idea that intentions are discrete states that directly cause action; suitable for readers with moderate background in neuroscience)

Ullmann-Margalit, E., & Morgenbesser, S. (1977). Picking and choosing. *Social Research, 44*(4), 757–785.

(Draws distinctions between picking (arbitrary actions) and choosing (deliberate ones) in relation to decisions and actions; suitable for all readers)

Van Boxtel, J., Tsuchiya, N., & Koch, C. (2010). Consciousness and attention: On sufficiency and necessity. *Frontiers in Psychology, 1*(217). doi:10.3389/fpsyg.2010.00217

(Summarizes psychophysical and neurophysiological evidence for a functional dissociation between top-down attention as analyzer and consciousness as synthesizer, and concludes that separating the effects of selective visual attention from those of visual consciousness is crucial to untangling their neural substrates; suitable for readers with moderate background in neuroscience)

van Gaal, S., Ridderinkhof, K. R., van den Wildenberg, W. P. M., & Lamme, V. A. F. (2009). Dissociating consciousness from inhibitory control: Evidence for unconsciously triggered response inhibition in the stop-signal task. *Journal of Experimental Psychology: Human Perception and Performance, 35*(4), 1129–1139.

(Uses masked (unconscious) or unmasked (conscious) stop-signals after participants are cued to act in order to demonstrate that some cognitive control functions can operate unconsciously; suitable for readers with moderate background in neuroscience)

Vargas, M. (2013). Situationism and moral responsibility: Free will in fragments. In T. Vierkant, J. Kiverstein, & A. Clark (Eds.), *Decomposing the Will* (pp. 325–350). Oxford University Press.

(Argues that theories of free will and moral responsibility that center on the ability of agents to respond to reasons must be modified, in light of situationist social psychology, to instead focus on conscious awareness of intentions; suitable for readers with minimal background in philosophy and psychology)

Verbaarschot, C., Farquhar, J., & Haselager, P. (2019). Free Wally: Where motor intentions meet reason and consequence. *Neuropsychologia, 133*, 107156.

(Using EEG and a probe method while subjects play an interactive deliberate decision-making game, times the what, when, and whether phases of intending; suitable for readers with minimal background in neuroscience)

Verschure, P. F. M. J. (2016). Synthetic consciousness: The distributed adaptive control perspective. *Philosophical Transactions of the Royal Society B: Biological Sciences, 371*(1701), 20150448. doi:doi:10.1098/rstb.2015.0448

(Presents the distributed adaptive control theory of consciousness as a way to answer the methodological and conceptual challenges that studying the nature of consciousness raises; suitable for readers with moderate background in neuroscience)

Vierkant, T. (2015). Is willpower just another way of tying oneself to the mast? *Review of Philosophy and Psychology, 6*(4), 779–790.

(Argues that free will is affected by one's ability to act in line with one's intentions and that one can resist limitations through effortful willpower, distraction, or structuring one's environment; suitable for readers with moderate background in philosophy)

Wegner, D. M. (2002). *The Illusion of Conscious Will* (Vol. 113). MIT Press.
(An empirically informed argument by a social psychologist that claims conscious will does not initiate action; suitable for all readers)

Wheaton, L. A., & Hallett, M. (2007). Ideomotor apraxia: A review. *Journal of the Neurological Sciences, 260*(1–2), 1–10.
(Reviews findings related to ideomotor apraxia (impaired ability to perform skilled gestures upon verbal command and/or by imitation) and its anatomical correlates; suitable for readers with minimal background in neuroscience)

Wiesenfeld, K., & Moss, F. (1995). Stochastic resonance and the benefits of noise: From ice ages to crayfish and SQUIDs. *Nature, 373*(6509), 33–36.
(Suggests that noise is useful for technological and biological systems because it enhances the detection of weak signals (stochastic resonance); suitable for readers with moderate background in neuroscience)

Wolf, S. (1990). *Freedom within Reason.* Oxford University Press.
(Charts a path between incompatibilism and compatibilism by arguing that freedom and responsibility require independence from forces that prevent or preclude us from choosing how to live in light of a sufficient appreciation of the world, but not from all forces beyond our control; suitable for readers with moderate background in philosophy)

Wolpert, D. M., & Kawato, M. (1998). Multiple paired forward and inverse models for motor control. *Neural Networks, 11*(7), 1317–1329. doi:https://doi.org/10.1016/S0893-6080(98)00066-5
(Advocates for a modular approach to motor learning and control and proposes a new architecture consisting of multiple paired forward–inverse models; suitable for readers with moderate background in neuroscience)

Zador, A. M. (2019). A critique of pure learning and what artificial neural networks can learn from animal brains. *Nature Communications, 10*(1), 3770. doi:10.1038/s41467-019-11786-6
(Argues that most animal behavior is not the result of clever learning algorithms but is rather encoded in the genome, which is the basis of animals' highly structured brain connectivity, which in turn enables rapid learning; suitable for readers with moderate background in neuroscience)

Zschorlich, V. R., & Köhling, R. (2013). How thoughts give rise to action-conscious motor intention increases the excitability of target-specific motor circuits. *PLoS One, 8*(12), e83845.
(Uses TMS to stimulate the motor cortices of human subjects forming intentions while also measuring muscle activity in their forearms to conclude that conscious intentions alter direction and force of movement; suitable for readers with moderate background in neuroscience)

Index

freedom from reasons-responsiveness
and, 35–38
God and, 37, 39
incompatibilism and, 33, 35, 42, 50–52,
55, 57, 59, 71, 75, 79, 92
libertarian theories of, 77–78
moral responsibility and, 34–39, 43,
84–85
neuroscience and, 37, 40
Free Wally (video game), 168
free will. *See also* freedom; will
artificial intelligence and, 80–81, 86–91
behavioral experiments regarding,
80–84
capacitarian accounts of, 60–62
computational models and, 266–74
consciousness and, 57–58, 61–62, 67,
73, 109–15, 217, 220
control and, 41, 47
deception and, 69
degrees of, 41–42, 45–46, 57–62,
70, 80
evolutionary perspectives on, 60
external influences as potential
obstacle to, 44–46, 57–59, 61
as folk concept, 82, 84
free action and, 65–67
genetics and, 84–85
historical dimension to, 44–46
intentions and, 57–59, 86, 89–90
introspection and, 83–85
libertarian free will and, 81
minimal and maximal thresholds for,
59–60, 62
moral responsibility and, 41, 46, 61,
119–26, 166
neuroscience and, 71–79, 122–24, 143
obligation and, 24, 68–70
probabilistic accounts of causation
and, 67–69
randomness and, 75–78, 92
rationality and, 46, 57–58, 61–62
readiness potential and, 74
semi-compatibilism and, 120–21
substantial self and, 81
Friedemann, Maja, 167
functional magnetic resonance imaging
(fMRI), 243, 247, 249–50

Gleason, C.A., 74
God, 37, 39
Granger causality, 247
Guillain Mollaret triangle, 177

Haggard, Patrick, 167–69, 266
Hallett, Mark, 252, 255–57
Hebbian cell assemblies, 269
Hobbes, Thomas, 37, 39
Hopfield network, 269, 272
Huxley, Aldous, 24, 30

Ia afferent nerve, 175–76
inferior frontal gyrus (IFG), 201
intentions
adaptive nature of, 8
artificial intelligence and, 86–88, 91
beliefs and, 89, 91, 206–7
Brentano-intentionality and, 91
Chinese-Room experiment
and, 87–88
computational models and, 268
conflicts between, 10–11
consciousness and, 9, 11–12, 111–15,
187, 189, 191–92, 217, 220
cued intentions and, 194–95, 198–99
decisions and, 95
deflationary intentions and, 114–15
desires compared to, 6–7, 9–11
direction of fit and, 5–10
distal intentions and, 8–9, 163–64,
185–88, 195–96, 251
free will and, 57–59, 86, 89–90
goals and, 185–91
inflationary intentions and, 114–15
intentional actions and, 11–12, 189
intentional binding and, 187
inverse modeling of, 186
as mental representations of the
future, 5–9
microfunctionalism and, 88
moral responsibility and, 125
motivation and, 195, 197, 200
motor intentions and, 8–10, 185, 188,
193–94, 196, 200
movements and, 251–55
neuroscience and, 185–200, 221, 254
physical actions and, 145, 148